朱利安諾的廚房 **06**

義大利麵

春Spring／夏Summer／秋Autumn／冬Winter

家鄉風味

朱利安諾‧格薩里（Giuliano Gasali）◎著

徐博宇‧廖家威◎攝影

Dear Readers,

I don't forget all of you. I come back to you with a new series of my cuisine books.

During the time I had been in Hong Kong, Vietnam, Singapore, Korea and several times China to make promotions and teach my style of cuisine.

When I was in Italy, I didn't stop to go around, to see, to observe and to research about the old cuisine (grandmother style). I can't wait to share my new discoveries with you.

Don't forget that the cuisine is an art and never finish to learn to improve your skill.

Bon Appetit!

Giuliano

〈作者序〉

親愛的讀者：

我並沒有遺忘你們，這一次，我帶著新系列的義大利食譜回來看你們。

過去這段時間，我去了香港、越南、新加坡、韓國，以及數次進出中國大陸，到亞洲各國宣傳與教授我的義大利料理。

而即使回到義大利時，我仍未曾停歇地四處尋覓、觀察與研究義大利傳統美食（祖母風味）。現在，我更是迫不及待地想要與你們分享我的心得。

別忘了，烹飪是一門藝術，需要永無止境的學習，方能擁有純熟的廚藝。

祝福你們

朱利安諾

<紀念版序>

拍攝這套食譜是在積木文化成立的那年。

轉眼十年流逝，主廚在完成拍攝工作，「鍋子高掛」幾年後，走入了歷史；晶華義大利餐廳cafe studio 也在2006年7月16日熄燈落幕。

朱利安諾擔任晶華義大利餐廳主廚期間，為許多文人饕客留下滿足的味蕾記憶；而當時與他一同工作的 夥伴們，紛飛之後，也憑藉著各自的天份和努力，散布出為數可觀的美味分子。

當年，主廚是在義大利完成這套書的校稿工作，這是他看過的唯一版本；而《朱利安諾的廚房》一問 市，馬上就成為積木的創社代表作。

源於對這個版本的深厚情感，因此，當決定發行十年紀念版時，我們決定仍保留當初的設計，只在封面 上加印銀色來顯現紀念版的意義。

白雲蒼狗，我們有幸能將朱利安諾畢生累積的廚藝編輯成書，傳承這套美味祕笈，如今回手更覺可貴。 感謝參與製作這套書的所有工作人員，包括攝影、審校、編輯們，紀念版則特別要感謝長年合作的製版 劉靜薏小姐、設計宜靜，因應檔案年代久遠、不堪使用所作的努力。

最後，我想代朱利安諾將此紀念版獻給他摯愛的妻子，沒有她的牽成，這套書就沒有誕生的可能。

積木文化總編輯　蔣豐雯

目錄

contents

百變義大利麵

1　貝殼麵 conchiglie
2　新鮮義式貓耳朵　fresh orechiette
3　紐卡第尼麵 gnochettini
4　雞冠麵 Creste di Gallo
5　卡瓦提利麵 cavatelli
6　義大利細扁麵 trenette
7　螺紋麵 gnocchette

8	季諾多麵	girandole
9	通心管麵	garganelli
10	波紋寬麵	reginette
11	通心麵	maccheroni
12	筆型麵	penne
13	布卡提尼麵	bucatini

8

9

10

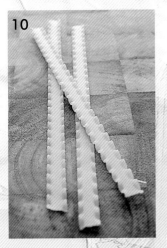

11

12

13

義大利麵的基本醬汁＆麵團

Pasta Dough 義大利麵團

材料：400公克

中筋麵粉 ·····················250公克
蛋 ·····························3個

Ingredients : 400 g

250 g all-purpose flour
3 eggs

作法

1. 將麵粉置於工作台上，在中間挖洞，放入打散的蛋液。
2. 用叉子慢慢地從中心處將麵粉與蛋和勻。
3. 攪拌均勻後，用雙手搓揉成形。
4. 持續揉麵約10分鐘，使之成為光滑有彈性的麵團。
5. 將麵團放入製麵機中，反覆壓平，使其成為薄紙般的麵皮，並可依需要加以變化。

a

b

Preparation

1. Pile the sifted flour onto the working surface and make a well in the center of the flour. In a small bowl lightly beat the eggs and pour it into the well.
2. With a fork, gradually draw in the flour from the inside wall of the well until the eggs are well integrated with the flour.
3. Use your hands to form a soft pasta.
4. With the heels of your hands, knead the dough for about 10 minutes until it has formed a ball that is smooth, elastic and not too hard.
5. Flatten the ball of dough through the roller. Repeat the procedure until the sheet of dough is paper thin. Cut it into the shape you need.

c

d

e

Colorful Pasta Dough 義大利麵調色盤

善用您的創意,只要在自製麵團中,加入五顏六色的蔬菜,就能幻化出百變的義大利麵食風情,享受創造的樂趣。

1 黃色義大利麵　　加入適量的番紅花粉。

2 粉紅色義大利麵　加入2大匙的紅甜菜泥(因紅甜菜在本地取材不易,可改成2大匙的熟胡蘿蔔泥代替,做成美麗的珊瑚麵)。

3 綠色義大利麵　　加入2大匙的菠菜泥。

4 紅色義大利麵　　加入1大匙的番茄糊。

5 黑色義大利麵　　加入1/2大匙的章魚墨囊。

1 **YELLOW**：Add a small amount of saffron.

2 **PINK**：Add 2 tablespoons of mashed beet root or carrot.

3 **GREEN**：Add 2 tablespoons of mashed spinach.

4 **RED**：Add 1 tablespoon of tomato paste.

5 **BLACK**：Add 1/2 tablespoon of cuttlefish ink.

Pesto Sauce 松子醬（青醬）

材料：350公克

松子·······························30公克
帕梅善起司粉···············50公克
九層塔·························100公克
特級純橄欖油···············150公克
大蒜切末·······················2瓣
鹽······························適量

Ingredients : 350 g

30 g pine nuts
50 g freshly grated parmesan cheese
100 g basil leaves
150 g extra-virgin olive oil
2 garlic cloves, finely chopped
a pinch of salt

作法

❶ 將松子、大蒜、橄欖油放入果汁機內打勻。

❷ 加入九層塔及鹽，持續攪打。

❸ 再放入帕梅善起司粉，續打數分鐘即成松子醬。可放入密封罐冷藏。

Preparation

❶ Puree the nuts and garlic with oil in a food processor.

❷ Add the basil leaves and salt. Continue to puree until a smooth emulsion is achieved.

❸ Add the Parmesan cheese, and puree for a few minutes. Place the sauce in a non-corrosive container. Keep the sauce refrigerated for ready to use.

a

b  c  d

Tomato Sauce 茄汁醬

材料：600公克

新鮮番茄	1.2公斤
大蒜	4瓣
九層塔切碎	30公克
特級純橄欖油	50公克
洋蔥切碎	70公克
鹽、胡椒	適量
辣椒切碎	適量

Ingredients : 600 g

1.2 kg ripe tomatoes
4 garlic cloves
30 g fresh basil leaves, coarsely chopped
50 g extra-virgin olive oil
70 g onions, finely chopped
salt and freshly ground pepper
fresh chili, chopped (optional)

作法

❶ 用刀在番茄末端劃個十字，放入沸水中煮約10秒鐘，取出剝皮、去籽、切碎。

❷ 橄欖油入鍋加熱，加入洋蔥及大蒜，炒約5分鐘。

❸ 再加入番茄煮至滾沸，以小火煮20至30分鐘，至醬汁濃稠後，加鹽、胡椒調味，酌量加入辣椒，熄火，加入九層塔，挑除大蒜即可。

Preparation

❶ Score an "X" on the base of each tomato. Immerse them in boiling water for few seconds and drain. Peel and cut crosswise in half. Squeeze the seeds and coarsely chop.

❷ Heat the olive oil in a saucepan, add the onions, garlic and sauté for 5 minutes, stirring frequently.

❸ Add the tomatoes and bring to the boil. Lower the heat and simmer for about 20-30 minutes until thickened. Season with salt, pepper and chili. Remove from the heat, add the chopped basil and discard the garlic.

Bechamel Sauce 鮮奶油醬

材料：1.1公斤

奶油……………………50公克
中筋麵粉…………………60公克
鹽………………………適量
豆蔻粉……………………適量
牛奶……………………1公升
奶油……………………1大匙

Ingredients : 1.1 kg

50 g butter
60 g all-purpose flour
salt
nutmeg
1 liter milk
1 tbsp butter

作法

❶ 奶油加熱融化，倒入麵粉用小火慢慢炒拌數分鐘（勿炒焦），待冷卻後備用。

❷ 牛奶煮沸倒入麵糊中，用力快速攪拌，然後轉小火邊煮邊攪拌。

❸ 至煮沸後，加鹽及豆蔻粉調味，然後過篩。

❹ 再刷上1大匙的奶油，以防止表層變硬。

Preparation

❶ Melt the butter in a non-corrosive saucepan. Add the flour and cook over low heat, stirring frequently for a few minutes (Do not brown the flour). Let it cool.

❷ Bring the milk to the boil. Pour the hot milk into the saucepan, whisking vigorously with a whisk.

❸ Simmer until the liquid boil, stirring frequently. Add the salt and nutmeg. Strain through a strainer to remove the lumps.

❹ Brush 1 tablespoon of butter to keep the skin tender.

a

b
c
d

Bolognese Sauce 義式肉醬

材料：1.5公斤

特級橄欖油	30公克
奶油	30公克
洋蔥切碎	80公克
胡蘿蔔切丁	80公克
芹菜切丁	80公克
大蒜切末	30公克
牛絞肉	500公克
豬絞肉	300公克
紅酒	2杯
罐頭番茄切碎	400公克
番茄糊	50公克
荷蘭芹切碎	50公克
鹽、胡椒	適量

Ingredients : 1.5 kg

30 g extra-virgin olive oil
30 g butter
80 g onions, finely chopped
80 g carrots, finely chopped
80 g celery, diced
30 g garlic, finely chopped
500 g beef, coarsely ground
300 g pork, coarsely ground
2 glass red wine
400 g canned tomatoes, coarsely chopped
50 g tomato paste
50 g parsley, finely chopped
salt and freshly ground pepper

作法

❶ 橄欖油、奶油入鍋加熱，炒香洋蔥、胡蘿蔔、芹菜和大蒜，以中火翻炒約5分鐘。

❷ 加入絞肉，拌炒約10分鐘。

❸ 倒入紅酒，煮至醬汁濃縮至一半。

❹ 接著放入番茄和番茄糊，小火續燉1小時，煮時仍須稍加攪拌，並撒些鹽、胡椒調味。

❺ 待醬料煮至濃稠，撒入荷蘭芹拌勻即可。

Preparation

❶ Heat the olive oil and butter in a large saucepan. Sauté the onions, carrots, celery and garlic over medium heat for 5 minutes.

❷ Add the ground meat and sauté for 10 minutes.

❸ Pour in the wine and cook until the liquid thickened.

❹ Add the tomatoes and tomato paste, and continue cooking over low heat for 1 hour. Season with salt and pepper, stirring occasionally.

❺ Cook until the sauce thickened. Add the parsley and stir well.

春
Spring

鍋子裡的節奏充滿煽惑

挑逗義大利麵翩翩起舞

擷取最鮮嫩的音符恣意躍動

春天的圓舞曲中有陽光灑落

春之蝶麵

材料　　4～6人份

蝴蝶麵	400公克
番茄	2個
四季豆	100公克
芹菜	80公克
節瓜	100公克
胡蘿蔔	100公克
大蒜切碎	1瓣
特級純橄欖油	50公克
鹽、胡椒	適量

Ingredients Serves: 4-6

400 g farfalle
2 ripe tomatoes
100 g French beans
80 g celery
100 g zucchini
100 g carrots
1 garlic clove, chopped
50 g extra-virgin olive oil
salt and freshly ground pepper

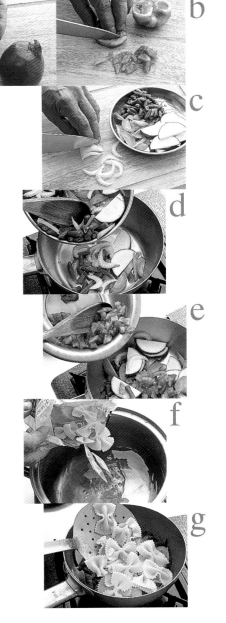

作法

❶ 番茄末端用刀劃出十字形，放入沸水中稍燙數秒鐘後取出，剝皮、去籽、切碎備用。

❷ 芹菜、節瓜、胡蘿蔔及四季豆切小塊。

❸ 鍋內倒入1/2的橄欖油加熱，先炒香大蒜，再放入芹菜、節瓜、胡蘿蔔及四季豆，拌炒數分鐘。

❹ 接著加入番茄，撒些鹽、胡椒調味，煮約3至4分鐘，煮時須稍加攪拌。

❺ 另將蝴蝶麵放入滾沸的鹽水中，煮10至12分鐘，至麵條軟硬適中。煮時須稍加攪動，以免沾鍋。

❻ 將蝴蝶麵撈出，倒入醬料中翻炒，加入剩餘的橄欖油續炒1分鐘，即可盛盤上桌。

Preparation

❶ Score an "X" on the base of each tomato. Immerse them in boiling water for few seconds and drain. Peel and cut crosswise in half. Squeeze the seeds and coarsely chop.

❷ Cut into thin slices the celery, zucchini, carrots and French beans.

❸ Heat half the olive oil in a frying pan and sauté the garlic for 1 minute. Add the French beans, celery, zucchini and carrots and sauté for few minutes.

❹ Add the tomatoes, salt and pepper, and cook for 3-4 minutes, stirring occasionally.

❺ Meanwhile bring the salted water to the boil. Add the farfalle and cook for 10-12 minutes until al dente, stirring occasionally.

❻ Drain the farfalle and toss with the sauce. Add the remaining olive oil, sauté for 1 minute and serve.

TRENETTE ALLA MODE DI LIPARI

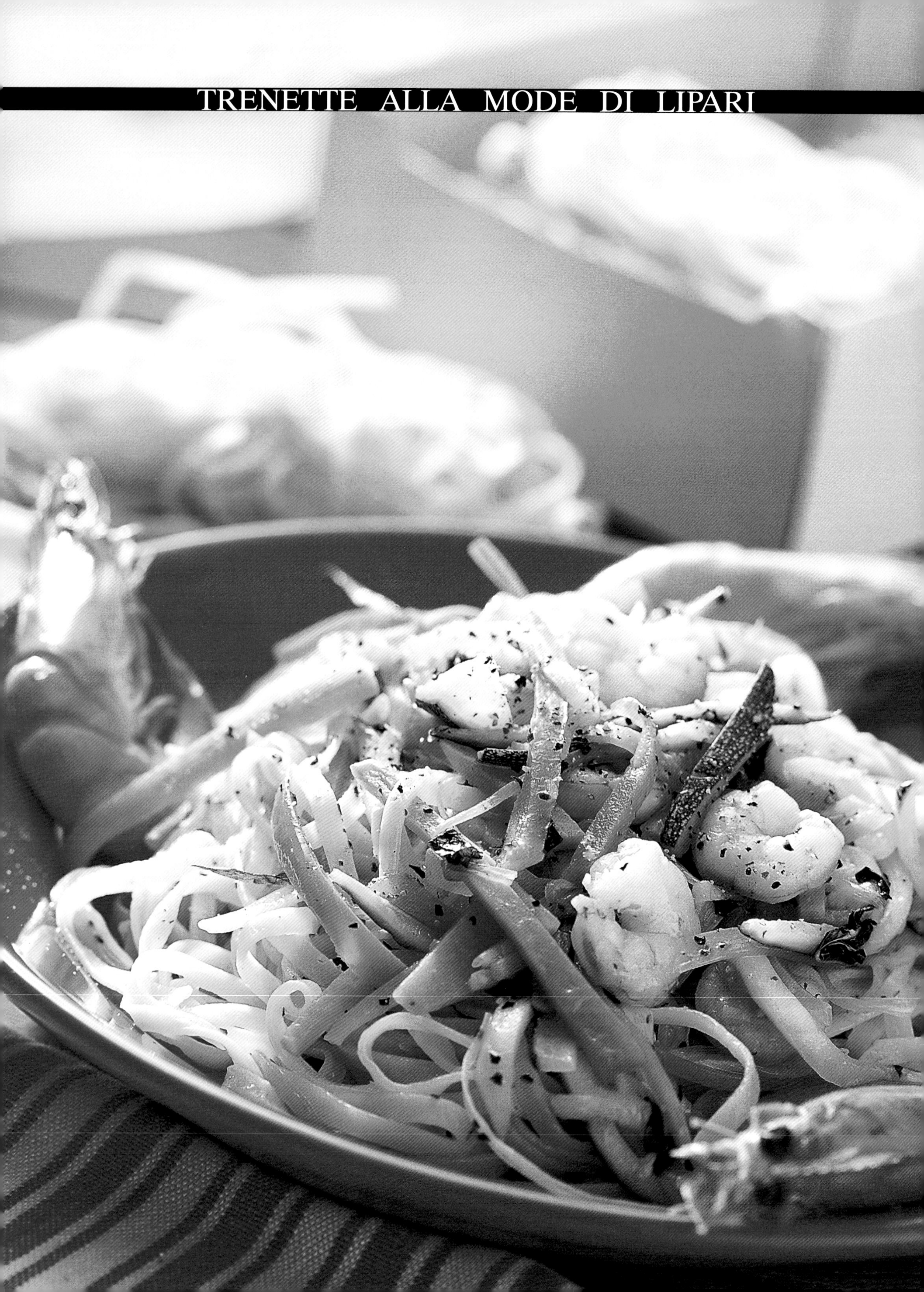

利巴瑞細麵

a
b
c
d
e
f

材料　　4～6人份

義大利細扁麵	400公克
節瓜	100公克
胡蘿蔔	100公克
蒜苗	100公克
九層塔	25公克
小蝦仁	400公克
白酒	1杯
奶油	20公克
特級純橄欖油	20公克
鹽、胡椒	適量

Ingredients Serves: 4-6

400 g trenette
100 g zucchini
100 g carrots
100 g leeks
25 g basil
400 g small shrimps
1 glass white wine
20 g butter
20 g extra-virgin olive oil
salt and freshly ground pepper

作法

❶ 胡蘿蔔、蒜苗、節瓜和九層塔切成細絲。

❷ 小蝦仁在冷水中洗淨後，用紙巾拭乾水分。

❸ 橄欖油和奶油放入平底鍋內加熱，放入所有蔬菜絲，炒6分鐘至蔬菜變軟，加入適量鹽、胡椒調味，稍加攪拌，再加入蝦仁同煮，倒入白酒，煮到酒精完全揮發。

❹ 另將麵條放入滾沸的鹽水中，煮6至7分鐘，至麵條軟硬適中。煮時須稍加攪動。

❺ 撈出麵條，與醬料翻炒1分鐘。

❻ 撒些現磨胡椒，即可趁熱盛盤上桌。

Preparation

❶ Cut into julienne pieces the carrots, leeks, zucchini and basil.

❷ Wash the shrimps under cold water and dry with kitchen paper.

❸ Heat the olive oil and butter in a frying pan. Add all the vegetables and sauté for 6 minutes until softened. Season with salt and pepper, stirring occasionally. Add the shrimps and white wine. Stir occasionally until the wine has evaporated.

❹ Meanwhile bring the salted water to the boil. Add the trenette and cook for 6-7 minutes until al dente, stirring occasionally.

❺ Drain the trenette and toss with the sauce. Sauté for 1 minute.

❻ Serve on individual warm dish. Sprinkle over the fresh pepper from pepper mill and serve.

藍佩度沙式蝶麵

材料　4～6人份

蝴蝶麵·····················400公克
淡菜肉·····················200公克
蝦仁·······················200公克
節瓜切片···················200公克
洋蔥切碎···················30公克
番茄切丁···················300公克
白酒·························1杯
荷蘭芹切碎·················30公克
特級純橄欖油···············40公克
鹽、胡椒···················適量

Ingredients Serves: 4-6

400 g farfalle
200 g shelled mussels
200 g peeled shrimps
200 g zucchini, sliced
30 g onions, finely chopped
300 g ripe tomatoes, diced
1 glass white wine
30 g parsley, finely chopped
40 g extra-virgin olive oil
salt and freshly ground pepper

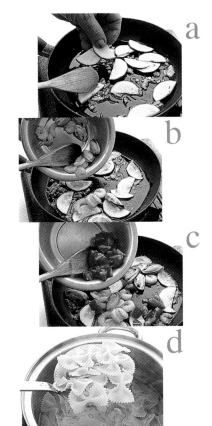

作法

❶ 橄欖油放入平底鍋內加熱，先炒香洋蔥，再加入節瓜炒軟。

❷ 接著加入淡菜、蝦仁及白酒，煮到酒精完全揮發，加適量鹽、胡椒調味。

❸ 再加入番茄續煮5分鐘。

❹ 另將麵條放入滾沸的鹽水中，煮12分鐘，至麵條軟硬適中。煮時須稍加攪動。

❺ 撈出麵條，與醬料翻炒均勻。

❻ 上桌前撒入荷蘭芹即可。

Preparation

❶ Heat the olive oil in a frying pan and sauté the onions for few minutes. Add the zucchini and cook until softened.

❷ Add the mussels, shrimps and wine and let to evaporate. Season with salt and pepper.

❸ Add the tomatoes and continue cooking for 5 minutes.

❹ Bring the salted water to the boil. Add the farfalle and cook for 12 minutes until al dente, stirring occasionally.

❺ Drain the farfalle and toss with the sauce.

❻ Sprinkle the chopped parsley over the pasta and serve.

CAVATELLI CON SEDANO E TONNO

鮪魚卡瓦提利

材料　　4～6人份

卡瓦提利麵	400公克
新鮮鮪魚	300公克
芹菜	200公克
洋蔥切碎	30公克
大蒜切碎	30公克
奶油	25公克
特級純橄欖油	30公克
檸檬皮屑	20公克
豆蔻粉	適量
白酒	1/2杯
鹽、胡椒	適量

Ingredients Serves: 4-6

400 g cavatelli
300 g fresh tuna fish
200 g celery
30 g onions, finely chopped
30 g garlic, finely chopped
25 g butter
30 g extra-virgin olive oil
20 g lemon rinds, grated
finely grated nutmeg
1/2 glass white wine
salt and freshly ground pepper

作法

❶ 鮪魚洗淨去皮，切成小塊。

❷ 芹菜切成薄片。

❸ 奶油和橄欖油放入平底鍋內加熱，炒香洋蔥和大蒜至變軟，加入芹菜拌炒，再倒入些許的熱水，煮約10分鐘。

❹ 加入鮪魚和白酒，續煮10分鐘，撒鹽、胡椒調味，加檸檬皮屑、豆蔻粉拌炒均勻。

❺ 另將麵條放入滾沸的鹽水中，煮12分鐘至麵條軟硬適中，煮時須稍加攪動，以免沾鍋。

❻ 撈出麵條，與醬料翻炒1分鐘，即可趁熱盛盤上桌。

Preparation

❶ Remove the skin from the tuna. Cut into small cubes.

❷ Slice the celery into thin pieces.

❸ Heat the butter and olive oil in a frying pan. Add the onions and garlic and sauté until softened. Add the celery and let to brown. Pour in some spoons of hot water and cook for 10 minutes.

❹ Add the tuna and wine and continue cooking for 10 minutes. Season with salt and pepper. Sprinkle the sauce with the lemon rinds and nutmeg.

❺ Meanwhile bring the salted water to the boil. Add the cavatelli and cook for 12 minutes until al dente, stirring occasionally.

❻ Drain the cavatelli and toss with the sauce. Sauté for 1 minute and serve immediately.

義式章魚麵

材料　　4～6人份

義大利麵 …………………400公克
章魚………………………600公克
大蒜切碎…………………25公克
洋蔥切碎…………………40公克
辣椒切碎…………………10公克
白酒 ………………………1杯
茄汁醬（見17頁）………200公克
荷蘭芹切碎 ………………30公克
特級純橄欖油 ……………40公克
鹽、胡椒…………………適量

Ingredients Serves: 4-6

400 g spaghetti
600 g octopus
25 g garlic, finely chopped
40 g onions, finely chopped
10 g fresh chili, finely chopped
1 glass white wine
200 g tomato sauce (see page 17)
30 g parsley, finely chopped
40 g extra-virgin olive oil
salt and freshly ground pepper

作法

❶ 章魚洗淨，切成小塊。

❷ 橄欖油放入平底鍋內加熱，炒香大蒜、洋蔥和辣椒，煮至洋蔥變軟。

❸ 放入章魚炒數分鐘，倒入白酒，煮到酒精完全揮發。

❹ 鍋內醬汁若太乾，可加些熱水調和，加蓋續煮15至20分鐘，撒些鹽、胡椒調味。

❺ 加入茄汁醬，續煮10分鐘。

❻ 另將麵條放入滾沸的鹽水中，煮約10分鐘，至麵條軟硬適中，煮時須稍加攪動。

❼ 撈出麵條，與醬料翻炒均勻，加入荷蘭芹拌炒1分鐘，即可趁熱上桌。

Preparation

❶ Clean and wash the octopus. Cut into pieces.

❷ Heat the olive oil in a frying pan. Add the garlic, onions and chili and sauté until the onions have softened.

❸ Add the octopus pieces and let to brown. Add the wine and allow to evaporate.

❹ If the sauce is too dry, add some hot water. Cover with the lid and cook for 15-20 minutes. Season with salt and pepper.

❺ Add the tomato sauce and continue cooking for 10 minutes.

❻ Meanwhile bring the salted water to the boil. Add the spaghetti and cook for 10 minutes until al dente, stirring occasionally.

❼ Drain the spaghetti and toss with the sauce. Add the parsley, sauté for 1 minute and serve.

蝦仁通心管麵

材料　　　4～6人份	Ingredients Serves: 4-6
通心管麵 ·················400公克	400 g garganelli
蝦 ·······················600公克	600 g shrimps
罐頭朝鮮薊 ·············300公克	300 g canned artichoke
特級純橄欖油 ···········40公克	40 g extra-virgin olive oil
紅蔥頭切碎 ·············30公克	30 g shallots, finely chopped
九層塔切碎 ·············30公克	30 g basil, finely chopped
白酒 ·······················1/2杯	1/2 glass white wine
鹽、胡椒 ·················適量	salt and freshly ground pepper

作法

❶ 將蝦剝殼去頭，在蝦背劃一刀，挑除腸泥，以冷水沖洗乾淨，再用紙巾拭乾水分。

❷ 朝鮮薊瀝乾油分，用湯匙剝成小塊。

❸ 橄欖油放入平底鍋內加熱，放入紅蔥頭炒數分鐘至變軟。

❹ 接著放入朝鮮薊和蝦仁，倒入白酒，撒鹽、胡椒調味，煮到酒精揮發。

❺ 另將麵條放入滾沸的鹽水中，煮約10分鐘，至麵條軟硬適中，煮時須稍加攪動。

❻ 撈出麵條，與醬料翻炒均勻，撒入九層塔拌炒1分鐘，趁熱盛盤上桌。

Preparation

❶ Peel the shrimps and remove the heads. Make a deep cut along the back of each shrimp and remove the dark intestinal vein. Rinse well under cold running water. Drain well and dry with kitchen paper.

❷ Drain the artichoke from the preserving oil and break with a spoon.

❸ Heat the olive oil in a frying pan. Add the shallots and sauté for few minutes until softened.

❹ Add the artichoke, shrimps, wine, salt and pepper. Stir until the wine has evaporated.

❺ Bring the salted water to the boil. Add the garganelli and cook for 10 minutes until al dente, stirring occasionally.

❻ Drain the garganelli and toss with the sauce. Sprinkle with the basil, sauté for 1 minute and serve.

巴利式貓耳朵

材料　　4～6人份	Ingredients Serves: 4-6
義式貓耳朵 ⋯⋯⋯⋯⋯400公克	400 g orechiette
綠花椰菜 ⋯⋯⋯⋯⋯⋯500公克	500 g broccoli
培根⋯⋯⋯⋯⋯⋯⋯150公克	150 g bacon
大蒜切碎⋯⋯⋯⋯⋯⋯25公克	25 g garlic, finely chopped
莫札雷拉起司 ⋯⋯⋯⋯150公克	150 g Mozzarella cheese
帕梅善起司粉 ⋯⋯⋯⋯40公克	40 g freshly grated Parmesan cheese
荷蘭芹切碎 ⋯⋯⋯⋯⋯40公克	40 g parsley, finely chopped
特級純橄欖油 ⋯⋯⋯⋯40公克	40 g extra-virgin olive oil
鹽、胡椒⋯⋯⋯⋯⋯⋯適量	salt and freshly ground pepper

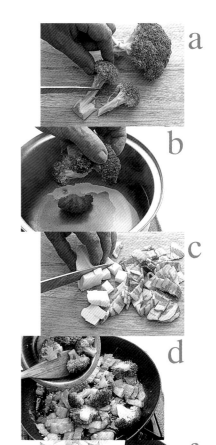

作法

1. 綠花椰菜洗淨，切成小朵。
2. 將綠花椰菜放入滾沸的鹽水中，煮8分鐘，瀝乾水分。鹽水保留備用。
3. 培根切成條狀。莫札雷拉起司切丁。
4. 平底鍋內倒入橄欖油加熱，炒香大蒜後，加入培根煮5分鐘，再加入綠花椰菜續炒5分鐘，撒些鹽、胡椒調味並攪拌均勻。
5. 另將麵條放入煮花椰菜的沸水中，煮10分鐘，撈出麵條，與醬料炒勻。
6. 加入莫札雷拉起司、帕梅善起司粉和荷蘭芹，拌炒1分鐘，即可盛盤上桌。

Preparation

1. Clean and wash the broccoli. Cut into florets.
2. Bring the salted water to the boil. Add the broccoli, cook for 8 minutes and drain. Reserve the salted water.
3. Cut the bacon into strips, and Mozzarella into cubes.
4. Heat the olive oil in a frying pan. Add the garlic and sauté for few minutes. Add the bacon and cook for 5 minutes. Add the broccoli and continue cooking for 5 minutes. Season with salt and pepper, stirring occasionally.
5. Add the orechiette in the same boiling water and cook for 10 minutes. Drain well and toss with the sauce.
6. Add the Mozzarella, Parmesan and parsley. Toss well, sauté for 1 minute and serve.

Three Colour Gnocchi with Leeks

彩繪洋芋麵

材料　　4～6人份

麵疙瘩部分：
馬鈴薯	1公斤
中筋麵粉	400公克
蛋	1個
橄欖油	1大匙
番茄糊	50公克
菠菜泥	50公克

醬料部分：
蒜苗切碎	300公克
胡蘿蔔切碎	80公克
芹菜切碎	80公克
奶油	30公克
特級純橄欖油	30公克
鮮奶油	100公克
帕梅善起司粉	40公克
鹽、胡椒	適量

Ingredients Serves: 4-6

For the gnocchi:
1 kg potatoes
400 g all-purpose flour
1 egg
1 tbsp olive oil
50 g tomato paste
50 g mashed spinach
For the sauce:
300 g leeks, chopped
80 g carrots, chopped
80 g celery, chopped
30 g butter
30 g extra-virgin olive oil
100 g whipping cream
40 g freshly grated Parmesan cheese
salt and freshly ground pepper

作法

❶ 馬鈴薯帶皮放入沸水中，煮約40分鐘，撈起待冷卻後，去皮搗成泥。

❷ 將馬鈴薯泥、蛋、麵粉及橄欖油混合均勻，搓揉成柔軟光滑的麵團。

❸ 將麵團分成三份，一份加入番茄糊，一份加入菠菜泥，另一份保持原色。分別將三份麵團搓揉成柔軟均勻的麵團。

❹ 取一小塊麵團搓成手指粗細的長條，切成2公分長的小段。

❺ 用叉子或其他適當器具將麵疙瘩壓出花紋，捲成環狀，撒些麵粉，以免相黏。

❻ 平底鍋內倒入奶油和橄欖油加熱，放入蒜苗、胡蘿蔔及芹菜，煮10分鐘，再倒入鮮奶油，加適量鹽、胡椒調味，煮時須稍加攪拌。

❼ 另將麵疙瘩放入滾沸的鹽水中，加蓋煮3至4分鐘，稍加攪拌，煮到麵疙瘩浮出水面。

❽ 撈出麵疙瘩，與醬料拌炒1分鐘，趁熱裝盤，撒帕梅善起司粉即可。

Preparation

❶ Cook the potatoes in boiling water for 40 minutes. Drain, let cool, peel and mash the potatoes.

❷ Knead the mashed potatoes with the egg, flour and olive oil gently, until the mixture is well mixed.

❸ Divide the mixture in three parts. Add tomato paste to one part, mashed spinach to another, and keep the other one as the original state. Knead the three parts until the mixture is well mixed.

❹ Take a handful of dough and roll into strips. Cut crosswise into 2 cm long pieces.

❺ Press each piece of dough to make the curve with a fork or any proper utensil, and dip in flour. There should be an indentation on one side and imprint of fork prongs on the other side. Roll into rings and keep each piece separated on the floured board.

❻ Simmer the butter and olive oil in a frying pan. Add the leeks, carrots and celery, and cook for 10 minutes. Add the cream, salt and pepper, stirring occasionally.

❼ Bring the salted water to the boil. Add the gnocchi, cover with a lid and cook for 3-4 minutes until they rise to the surface, stirring occasionally.

❽ Drain the gnocchi and toss with the sauce. Sauté for 1 minute. Serve immediately on a serving dish or individual plate, and sprinkle with Parmesan cheese.

37

海鮮卡彭那拉

材料 4～6人份		Ingredients Serves: 4-6
寬扁麵	400公克	400 g tagliolini
紅蔥頭切碎	40公克	40 g shallots, finely chopped
蛤蜊肉	200公克	200 g shelled clams
小透抽	250公克	250 g small squids
蝦仁	200公克	200 g peeled shrimps
新鮮鮭魚	200公克	200 g fresh salmon
白酒	1杯	1 glass white wine
蛋黃	2個	2 egg yolks
鮮奶油	200公克	200 g whipping cream
特級純橄欖油	40公克	40 g extra-virgin olive oil
荷蘭芹切碎	30公克	30 g parsley, finely chopped
鹽、胡椒	適量	salt and freshly ground pepper

作法

1 透抽洗淨，清除內臟，切條。

2 鮭魚去皮，切丁。

3 用刀劃開蝦背，挑除腸泥，用水沖淨，再用紙巾吸乾水分。

4 橄欖油放入平底鍋內加熱，炒香紅蔥頭至變軟，放入透抽炒5分鐘，並稍加攪拌。

5 接著加入蛤蜊肉、鮭魚和蝦仁，續煮3至4分鐘，撒鹽、胡椒調味，倒入白酒煮到酒精完全揮發。

6 用一小碗，將蛋黃、鮮奶油、鹽、胡椒及荷蘭芹拌勻。

7 另將麵條放入滾沸的鹽水中，煮6至7分鐘至麵條軟硬適中，撈出與海鮮醬料翻炒均勻。

8 最後倒入拌勻的蛋黃醬汁，快速攪拌，即可趁熱盛盤上桌。

Preparation

1 Remove the inner organs from the squids and cut into strips.

2 Remove the skin from the salmon and cut into cubes.

3 Make a deep cut along the back of each peeled shrimp and remove the dark intestinal vein. Rinse well under cold running water. Drain well and dry with kitchen paper.

4 Heat the olive oil in a frying pan. Add the shallots and sauté until softened. Add the squids and cook for 5 minutes, stirring constantly.

5 Add the clams, salmon and shrimps and cook for 3-4 minutes. Season with salt and pepper. Pour in the wine and stir well until it has evaporated.

6 In a bowl, whisk the egg yolks, cream, salt, pepper and parsley.

7 Bring the salted water to the boil. Add the tagliolini and cook for 6-7 minutes until al dente. Drain and toss with the sauce.

8 Pour in the egg mixture over the tagliolini, stir quickly and serve on a warm serving dish.

朝鮮薊千層麵

材料　　4～6人份	Ingredients Serves: 4-6
義大利麵團（見14頁）…600公克	600 g pasta dough (see page 14)
鮮奶油醬（見18頁）……400公克	400 g Bechamel Sauce (see page 18)
罐頭朝鮮薊瀝乾、切條…600公克	600 g canned artichoke, drained and cut in strips
大蒜切碎………………………30公克	30 g garlic, finely chopped
荷蘭芹切碎……………………30公克	30 g parsley, finely chopped
帕梅善起司粉……………100公克	100 g freshly grated Parmesan cheese
奶油………………………………80公克	80 g butter
蛋黃………………………………3個	3 egg yolks
鹽、胡椒………………………適量	salt and freshly ground pepper

作法

❶ 平底鍋內倒入1/2奶油加熱，放入大蒜和荷蘭芹，炒約3至4分鐘，加入朝鮮薊，撒適量鹽、胡椒調味，續炒5分鐘。

❷ 在碗內放入炒過的朝鮮薊，加入蛋黃、鮮奶油醬和2/3的帕梅善起司粉，用湯匙拌勻。

❸ 將麵團擀成薄片，再切成8×12公分的麵皮數片。

❹ 將麵皮放入滾沸的鹽水中，煮4分鐘，每次勿煮超過7片，並須稍加攪動。煮好後，用漏杓撈起，放入冷水浸泡後再瀝乾，一片片平放在略濕的布巾上，並以保鮮膜覆蓋，以免麵皮變乾。

❺ 烤盤刷上奶油，放一層麵皮，再鋪放朝鮮薊醬料。重複疊放麵皮及醬料的步驟，直到材料用完。確定最上層是朝鮮薊醬料，再撒些帕梅善起司粉，放幾塊奶油。

❻ 放進預熱至200℃（400℉）的烤箱中，烤25至30分鐘，直到表面金黃，取出放置5分鐘後，再盛盤上桌。

Preparation

❶ Heat half the butter in a frying pan. Add the garlic and parsley and sauté for 3-4 minutes. Add the artichoke, salt and pepper and sauté for 5 minutes, stirring occasionally.

❷ Put the artichoke in a bowl. Add the egg yolks, Bechamel and 2/3 of Parmesan cheese. Mix well.

❸ Roll out the dough into a thin sheet and cut into 8×12㎝ pieces.

❹ Bring the salted water to the boil. Add the lasagna and cook for 4 minutes, stirring occasionally. Do not cook more than 7 pieces at a time. Drain with a sieve. Put in cold water and drain again. Arrange the lasagna one by one on the tray, lined with dampened towel and cover with plastic wrap.

❺ Butter the baking dish. Drape the lasagna and spread the mixture. Repeat the procedure until all the ingredients are used. Make sure the final layer is the sauce, topped with remaining Parmesan cheese and some pieces of butter.

❻ Heat the oven to 200℃(400℉) and bake the lasagna for 25-30 minutes until the top is golden browned. Remove the lasagna, let rest for 5 minutes and serve.

蔬菜千層翡翠麵

材料　　4～6人份

節瓜…………………………300公克	
茄子…………………………300公克	
甜椒…………………………300公克	
鮮奶油醬（見18頁）……300公克	
特級純橄欖油……………100公克	
菠菜麵團（見15頁）……600公克	
九層塔切碎………………30公克	
莫札雷拉起司切丁………150公克	
奶油…………………………60公克	
鹽、胡椒…………………適量	

Ingredients Serves: 4-6

300 g zucchini
300 g eggplants
300 g bell peppers
300 g Bechamel sauce (see page 18)
100 g extra-virgin olive oil
600 g spinach pasta dough (see page 15)
30 g basil, chopped
150 g Mozzarella Cheese, diced
60 g butter
salt and freshly ground pepper

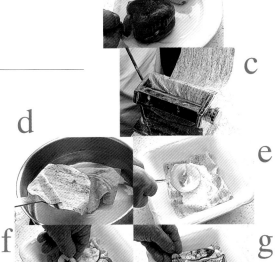

作法

❶ 節瓜和茄子洗淨，切成厚0.5公分的片狀，淋些橄欖油，撒鹽、胡椒，放在烤爐上，烤至表面微黃。

❷ 將甜椒以大火烤至外皮變黑，烤時須持續翻面。待甜椒冷卻後，去皮並縱向對切，切除中心的核及籽。

❸ 將麵團擀成薄片，切成8×12公分的麵皮數片。

❹ 將麵皮放入滾沸的鹽水中，煮4分鐘，每次勿煮超過7片，煮時須稍加攪動。煮好用漏杓撈起，放入冷水浸泡再瀝乾，一片片平放在略濕的布巾上，並以保鮮膜覆蓋，以免麵皮變乾。

❺ 烤盤刷上奶油，放一層麵皮，塗抹鮮奶油醬，接著鋪一層節瓜，撒些九層塔和莫札雷拉起司。

❻ 重複上述疊放麵皮、醬料的步驟，依序鋪擺節瓜一層、茄子一層、甜椒一層。確定最上層是鮮奶油醬，再撒入莫札雷拉起司和數塊奶油。

❼ 放進預熱至200℃（400°F）的烤箱中，烤25至30分鐘，直到表面金黃即可。

Preparation

❶ Wash and clean the zucchini and eggplants, and cut lengthwise into 0.5cm thick slices. Sprinkle with olive oil, salt and pepper and grill until lightly browned.

❷ Bake the bell peppers over high heat, turning constantly until the skin has blackened completely. Let the bell peppers cool, peel and cut in half lengthwise. Remove the cores and discard the seeds.

❸ Roll out the pasta dough into a thin sheet and cut into 8×12cm pieces.

❹ Bring the salted water to the boil. Add the lasagne and cook for 4 minutes. Do not cook more than 7 pieces at a time. Drain with a sieve. Put in cold water and drain again. Arrange the lasagna one by one on trays, lined with dampened towels and cover with plastic wrap.

❺ Butter the baking dish. Drape the lasagna and spread with some Bechamel. Cover with the zucchini, and spread with basil and Mozzarella.

❻ Repeat the procedure, and arrange the zucchini, eggplants and bell peppers in order. Make sure the final layer is the Bechamel, topped with Mozzarella cheese and some pieces of butter.

❼ Heat the oven to 200℃(400°F) and bake the lasagna for 25-30 minutes until the top is golden browned.

蔬菜醬魚餃

材料　　4～6人份

| 義大利麵團（見14頁）…400公克 |
| 餡料部分： |
| 白色魚肉，燙熟 …………300公克 |
| 白麵包切丁 ………………100公克 |
| 蛋 ……………………………1個 |
| 鮮奶油……………………80公克 |
| 百里香……………………10公克 |
| 牛膝草……………………10公克 |
| 醬料部分： |
| 洋蔥切碎……………………40公克 |
| 黑橄欖切碎 ………………120公克 |
| 酸豆………………………40公克 |
| 百里香……………………10公克 |
| 牛膝草……………………10公克 |
| 節瓜切丁 …………………100公克 |
| 番茄切丁 …………………100公克 |
| 黃椒切丁 …………………100公克 |
| 奶油………………………30公克 |
| 鹽、胡椒……………………適量 |

Ingredients Serves: 4-6

400 g pasta dough (see page 14)
For the stuffing:
300 g fish (white meat), steamed
100 g white bread, diced
1 egg
80 g whipping cream
10 g thyme
10 g marjoram
For the sauce:
40 g onions, chopped
120 g black olives, coarsely chopped
40 g capers
10 g thyme
10 g marjoram
100 g zucchini, diced
100 g ripe tomatoes, diced
100 g yellow bell peppers, diced
30 g butter
salt and freshly ground pepper

作法

❶ 將魚肉、麵包、蛋、鮮奶油、百里香、牛膝草、鹽和胡椒拌勻。

❷ 取一塊麵團擀成6至7公分寬的麵皮，在四周刷些水以便沾黏，以直線排列的方式，每隔3公分放一大匙（作法1）的餡料，再將麵皮對折包覆餡料。

❸ 壓出空氣，用滾輪刀將麵皮切成3公分見方的正方形，四周壓緊黏住。

❹ 鍋內加熱奶油，炒香洋蔥至變軟，放入橄欖、酸豆、蔬菜、香料、鹽和胡椒，煮10分鐘。拌炒時可視情況加些熱水調和。

❺ 另將魚餃放入滾沸的鹽水中，煮6至8分鐘，稍加攪動，待熟後用漏杓撈起。

❻ 將魚餃放入醬料中，翻炒約1分鐘，即可盛盤上桌。

Preparation

❶ Pour the fish in a bowl. Add the bread, egg, cream, thyme, marjoram, salt and pepper. Mix well.

❷ Roll out a strip of pasta dough into 6-7㎝ wide sheet. Brush the edges of each strip with water for sticking. Place a tablespoon of fish mixture at 3㎝ intervals along the sheet. Cover with the second pasta sheet.

❸ Press out the air pockets. With a pastry wheel, cut the pasta into 3×3㎝ squares. Press around each parcel to seal the edges.

❹ Heat the butter in a saucepan. Add the onions and sauté until soft. Add the olives, capers, all the vegetables, herbs, salt and pepper. Sauté for 10 minutes. If necessary, pour some hot water, stirring occasionally.

❺ Bring the salted water to the boil. Add the ravioli and cook for 6-8 minutes until al dente, stirring occasionally. Drain with a sieve.

❻ Toss the ravioli with the sauce and sauté for 1 minute. Serve immediately.

45

夏
Summer

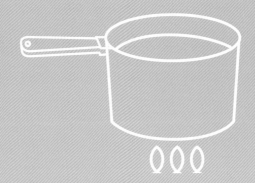

乘著夢想之翼迎風翱翔

划著希望的槳破浪而行

輕點魔法棒施展美味咒語

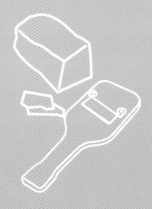

義大利麵的世界充滿無限可能

卡拉巴利式麵

材料　　4～6人份

卡瑟瑞雀麵	400公克
莫札雷拉起司切丁	150公克
紅椒切細絲	1個
黃椒切細絲	1個
大蒜切碎	20公克
大蒜	1瓣
酸豆	50公克
綠橄欖	80公克
香菇（貯存於油中）	80公克
鯷魚	50公克
乾燥俄力岡	20公克
番茄切碎	250公克
特級純橄欖油	60公克
鹽、胡椒	適量

Ingredients Serves: 4-6

- 400 g casereccie
- 150 g Mozzarella cheese, diced
- 1 red bell pepper, cut in juliennes
- 1 yellow bell pepper, cut in juliennes
- 20 g garlic, coarsely chopped
- 1 garlic clove
- 50 g capers
- 80 g green olives
- 80 g mushrooms, preserved in oil
- 50 g anchovy
- 20 g dry oregano
- 250 g ripe tomatoes, coarsely chopped
- 60 g extra-virgin olive oil
- salt and freshly ground pepper

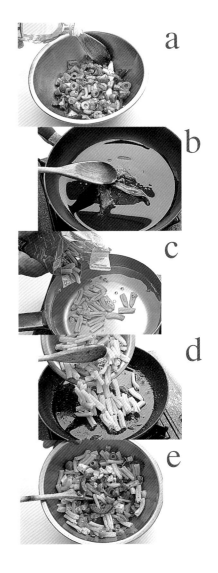

作法

❶ 將一瓣大蒜放入沙拉碗內搗碎，加入紅椒、黃椒、酸豆、橄欖、莫札雷拉起司、俄力岡香料、香菇，以及切碎的番茄、大蒜，倒入1/2橄欖油拌勻，醃漬1小時使其入味。

❷ 平底鍋內倒入剩餘的橄欖油加熱，放入鯷魚炒至融化。

❸ 另將麵條放入滾沸的鹽水中，煮12分鐘至麵條軟硬適中，煮時須稍加攪動，以免沾鍋。

❹ 撈出麵條，放入平底鍋內與鯷魚翻炒。

❺ 將炒過的麵條與（作法1）的醬料拌勻，加適量鹽、胡椒調味，即可上桌。

Preparation

❶ Crush the garlic clove in a salad bowl. Add the bell peppers, capers, olives, Mozzarella cheese, oregano, mushrooms, chopped garlic and chopped tomatoes. Dress with half the olive oil and let to marinate for 1 hour.

❷ Heat the remaining olive oil in a frying pan, add the anchovy and let to melt.

❸ Meanwhile bring the salted water to the boil. Add the casereccie and cook for 12 minutes until al dente, stirring occasionally.

❹ Drain the casereccie and toss with the melted anchovy.

❺ Toss the pasta with the cooled sauce, adjust the salt and pepper, and serve immediately.

貝沙利瑞式麵條

材料　　4～6人份	Ingredients Serves: 4-6
布卡提尼麵 ⋯⋯⋯⋯⋯⋯400公克	400 g bucatini
培根⋯⋯⋯⋯⋯⋯⋯⋯150公克	150 g bacon
火腿⋯⋯⋯⋯⋯⋯⋯⋯150公克	150 g ham
茄汁醬（見17頁）⋯⋯⋯400公克	400 g tomato sauce (see page 17)
洋蔥切片⋯⋯⋯⋯⋯⋯⋯40公克	40 g onions, sliced
大蒜切碎⋯⋯⋯⋯⋯⋯⋯20公克	20 g garlic, finely chopped
帕梅善起司粉 ⋯⋯⋯⋯⋯50公克	50 g freshly grated Parmesan cheese
荷蘭芹切碎⋯⋯⋯⋯⋯⋯30公克	30 g parsley, coarsely chopped
辣椒切碎⋯⋯⋯⋯⋯⋯⋯10公克	10 g fresh chili, coarsely chopped
特級純橄欖油 ⋯⋯⋯⋯⋯40公克	40 g extra-virgin olive oil
奶油⋯⋯⋯⋯⋯⋯⋯⋯⋯20公克	20 g butter
白酒 ⋯⋯⋯⋯⋯⋯⋯⋯⋯1/2杯	1/2 glass white wine
鹽 ⋯⋯⋯⋯⋯⋯⋯⋯⋯⋯適量	salt

作法

❶ 培根、火腿切成條狀。

❷ 橄欖油和奶油放入平底鍋內加熱，炒香洋蔥、大蒜和辣椒，煮到洋蔥變軟，放入培根、火腿炒8分鐘，倒入白酒，煮到酒精完全揮發。

❸ 加入茄汁醬，續煮10分鐘。

❹ 另將麵條放入滾沸的鹽水中，煮12分鐘至軟硬適中，煮時須稍加攪拌。

❺ 撈出麵條，與醬料翻炒1分鐘。

❻ 趁熱盛盤，撒入荷蘭芹和帕梅善起司粉，即可上桌。

Preparation

❶ Cut the bacon and ham into strips.

❷ Heat the olive oil and butter in a frying pan. Add the chili, onions and garlic, and cook until the onions have softened. Add the bacon and ham and sauté for 8 minutes. Pour in the wine and allow to evaporate.

❸ Add the tomato sauce and continue cooking for 10 minutes.

❹ Bring the salted water to the boil. Add the bucatini and cook for 12 minutes until al dente, stirring occasionally.

❺ Drain the bucatini, toss with the sauce and sauté for 1 minute.

❻ Serve on a warm serving dish. Sprinkle over the parsley and Parmesan cheese.

夏季之麵

材料　　4～6人份	Ingredients Serves: 4-6
車輪麵……………………400公克	400 g ruote
豬油……………………100公克	100 g pork fat
荷蘭芹…………………30公克	30 g parsley
大蒜………………………2瓣	2 garlic cloves
番茄……………………400公克	400 g ripe tomatoes
九層塔切碎………………30公克	30 g basil, chopped
帕梅善起司粉……………60公克	60 g freshly grated Parmesan cheese
特級純橄欖油……………20公克	20 g extra-virgin olive oil
鹽、胡椒…………………適量	salt and freshly ground pepper

作法

❶ 豬油、大蒜及荷蘭芹混合剁碎。

❷ 番茄末端用刀劃個十字，放入滾沸的熱水中燙數秒鐘，撈起，剝除外皮，去籽後切丁。

❸ 將橄欖油和（作法1）的豬油放入平底鍋內加熱，放入番茄煮10分鐘，撒些鹽、胡椒調味，煮時須稍加攪拌。

❹ 另將麵條放入滾沸的鹽水中，煮12分鐘至麵條軟硬適中，煮時須稍加攪動，以免沾鍋。

❺ 撈出麵條，與醬料翻炒均勻，加入九層塔，再拌炒約1分鐘。

❻ 盛盤上桌前，撒入帕梅善起司粉及現磨胡椒即可。

Preparation

❶ With a big knife, chop finely the pork fat, garlic and parsley.

❷ Score an "X" on the base of tomatoes. Immerse them in boiling water for few seconds and drain. Peel and cut crosswise in half. Squeeze the seeds and cut into cubes.

❸ Heat the olive oil and pork fat mixture in a frying pan. Add the tomato cubes and cook for 10 minutes. Season with salt and pepper, stirring occasionally.

❹ Bring the salted water to the boil. Add the ruote and cook for 12 minutes until al dente, stirring occasionally.

❺ Drain the ruote and toss with the sauce. Add the basil and sauté for 1 minute.

❻ Serve on a warm serving dish. Sprinkle over the Parmesan cheese and fresh pepper from the pepper mill.

風尚通心麵

材料　　4～6人份

通心麵·····················400公克
大番茄切片、保留汁液·····1.2公斤
洋蔥切薄片··················200公克
黑橄欖去籽·················120公克
酸豆·······················40公克
新鮮俄力岡·················20公克
特級純橄欖油···············60公克
鹽、胡椒····················適量

Ingredients Serves: 4-6

400 g maccheroni
1.2 kg big tomatoes, sliced and reserve the juice
200 g onions, thinly sliced
120 g black olives, seeded
40 g capers
20 g fresh oregano
60 g extra-virgin olive oil
salt and freshly ground pepper

a
b
c
d
e
f
g

作法

❶ 在塗油的耐熱盤內鋪擺1/2的番茄片，加適量鹽、胡椒調味，撒入1/2的俄力岡。

❷ 再將1/2的洋蔥片平均鋪擺在番茄之上，淋入2大匙橄欖油。

❸ 另將麵條放入滾沸的鹽水中，煮6分鐘至麵條軟硬適中，煮時須稍加攪動，以免沾鍋。

❹ 撈出麵條放入大碗中，與番茄汁、橄欖、酸豆、胡椒、些許俄力岡及2大匙橄欖油拌勻。

❺ 將拌好的麵條均勻地放在洋蔥片上。

❻ 在麵條上鋪擺剩餘的番茄、洋蔥和俄力岡，淋入橄欖油，放入預熱180℃（350℉）的烤箱內，烤10至15分鐘，取出待數分鐘後，即可盛盤上桌。

Preparation

❶ Grease a heat-resistant pan, pour half the tomatoes, salt and pepper. Sprinkle over half the oregano.

❷ Pour over half the onions and 2 tablespoons of olive oil.

❸ Bring the salted water to the boil. Add the maccheroni and cook for 6 minutes until al dente, stirring occasionally.

❹ Drain the maccheroni. Pour in a bowl and dress with the tomato juice, olives, capers, pepper, some oregano and 2 tablespoons of olive oil. Mix well.

❺ Spoon the pasta over the onions.

❻ Cover the pasta with the remaining tomatoes, onions and oregano. Spoon over some olive oil. Cook in a preheated oven of 180℃(350℉) for 10-15 minutes. Remove from the oven, allow to rest for few minutes and serve.

海中鮮麵

材料　　4～6人份

新鮮義式貓耳朵 ··········· 300公克
明蝦 ·········· 500公克
番茄 ·········· 500公克
地中海豆煮熟 ·········· 250公克
洋蔥切碎 ·········· 60公克
白酒 ·········· 1杯
鼠尾草葉切碎 ·········· 10片
特級純橄欖油 ·········· 50公克
鹽、胡椒 ·········· 適量

Ingredients Serves: 4-6

300 g fresh orechiette
500 g prawns
500 g ripe tomatoes
250 g chick-peas, boiled
60 g onions, finely chopped
1 glass white wine
10 sage leaves, chopped
50 g extra-virgin olive oil
salt and freshly ground pepper

作法

❶ 番茄末端用刀劃個十字，放入滾沸的熱水中燙數秒鐘，撈起後，剝皮，去籽切碎。

❷ 橄欖油放入平底鍋內加熱，炒香洋蔥，加入鼠尾草煮5分鐘。

❸ 接著加入明蝦，用大火炒數分鐘，倒入白酒，待酒精揮發後，續煮2至3分鐘。

❹ 取出明蝦，待冷卻後剝殼，但保留頭尾。

❺ 同鍋內加入地中海豆、2至3大匙的水和切碎的番茄，煮約5分鐘，將明蝦放回鍋內，撒些鹽、胡椒調味。

❻ 另將貓耳朵放入滾沸的鹽水中，煮10至12分鐘至軟硬適中，須稍加攪動，以免沾鍋。

❼ 將貓耳朵撈起瀝乾，倒入醬料中翻炒均勻，趁熱盛盤，並挑出明蝦擺在麵上裝飾即可。

Preparation

❶ Score an "X" on the base of each tomato. Immerse them in boiling water for few seconds. Drain, peel and cut crosswise in half. Squeeze the seeds and finely chop.

❷ Heat the olive oil in a frying pan. Sauté the chopped onions until softened. Add the sage leaves and cook for 5 minutes.

❸ Add the prawns and sauté over high heat. Add the white wine and cook until the wine has evaporated. Continue cooking for 2-3 minutes.

❹ Drain the prawns and let cool. Remove the shell. Do not remove the head and tail.

❺ Put the chick-peas in the frying pan with 2-3 tablespoons of water and the chopped tomatoes, and cook for 5 minutes. Put back the prawns. Adjust the seasoning with salt and pepper.

❻ Meanwhile bring the salted water to the boil. Add the orechiette and cook for 10-12 minutes until al dente, stirring occasionally.

❼ Drain the orechiette and toss with the sauce. Serve hot and use the prawns to decorate.

SPAGHETTI ALLA VESUVIO

火山式義大利麵

材料	4～6人份
義大利麵	400公克
新鮮鯷魚	500公克
洋蔥切碎	20公克
大蒜切碎	20公克
荷蘭芹切碎	30公克
九層塔切碎	20公克
松子	30公克
白酒	1/2杯
茄汁醬（見17頁）	400公克
特級純橄欖油	30公克
九層塔葉	少許
鹽、胡椒	適量

Ingredients Serves: 4-6

400 g spaghetti
500 g fresh anchovy
20 g onions, finely chopped
20 g garlic, finely chopped
30 g parsley, finely chopped
20 g basil, finely chopped
30 g pine nuts
1/2 glass white wine
400 g tomato sauce (see page 17)
30 g extra-virgin olive oil
basil leaves
salt and freshly ground pepper

作法

❶ 清除鯷魚內臟、頭及魚骨，洗淨，切碎。

❷ 將橄欖油放入平底鍋內加熱，炒香洋蔥、大蒜、荷蘭芹及九層塔，煮約數分鐘。

❸ 加入鯷魚，用木杓輕輕翻炒，煮約2分鐘後，再倒入白酒，煮到酒精完全揮發。

❹ 接著加入茄汁醬和松子，加適量鹽、胡椒調味，加蓋以小火煮10分鐘。

❺ 另將麵條放入滾沸的鹽水中，煮10至12分鐘，至麵條軟硬適中，煮時須稍加攪動。

❻ 撈出麵條，與醬料翻炒均勻，趁熱盛盤，擺些九層塔葉裝飾，即可上桌。

Preparation

❶ Clean the anchovy. Remove the head, bone and intestine. Chop the anchovy coarsely.

❷ Heat the olive oil in a frying pan. Add the onions, garlic, parsley and basil, and sauté for few minutes.

❸ Add the anchovy and sauté for 2 minutes, tossing gently with a wooden spoon. Add the wine and cook until it has evaporated.

❹ Add the tomato sauce and pine nuts, and season with salt and pepper. Cover with a lid and simmer for 10 minutes.

❺ Meanwhile bring the salted water to the boil. Cook the spaghetti for 10-12 minutes until al dente, stirring occasionally.

❻ Drain the spaghetti, toss with the sauce and serve on a warm serving dish with some basil leaves for decoration.

鮪魚義大利麵

材料　　4～6人份

義大利麵 ·······················400公克
乾燥義式香菇 ··············50公克
大蒜切碎·······················20公克
番茄糊···························80公克
罐頭鮪魚 ·····················250公克
荷蘭芹切碎 ··················30公克
奶油·····························20公克
特級純橄欖油 ··············30公克
鹽、胡椒·························適量

Ingredients Serves: 4-6

400 g spaghetti
50 g dry mushrooms (cepes)
20 g garlic, finely chopped
80 g tomato paste
250 g canned tuna
30 g parsley, finely chopped
20 g butter
30 g extra-virgin olive oil
salt and freshly ground pepper

作法

❶ 將義式香菇浸泡於溫水中10分鐘，瀝乾水分，切碎備用。

❷ 番茄糊倒入碗中，加些許溫水稀釋。

❸ 將奶油、橄欖油放入平底鍋內加熱，先放入大蒜炒香，再加入香菇同炒數分鐘。

❹ 倒入稀釋的番茄糊，用小火燉煮25分鐘，撒適量鹽、胡椒調味，煮時須稍加攪拌。

❺ 再加入瀝乾油分的鮪魚，同煮5分鐘。

❻ 另將麵條放入滾沸的鹽水中，煮約7分鐘至麵條軟硬適中，煮時須稍加攪動，以免沾鍋。

❼ 撈出麵條，與醬料翻炒均勻，撒些荷蘭芹再拌炒1分鐘，即可盛盤上桌。

Preparation

❶ Keep the dry mushrooms in warm water for 10 minutes until softened. Drain and coarsely chop.

❷ Pour the tomato paste in a bowl. Add some warm water to dilute.

❸ Heat the butter and olive oil in a frying pan. Add the garlic and sauté for few minutes. Add the mushrooms and brown for few minutes.

❹ Add the diluted tomato paste and cook over low heat for 25 minutes. Season with salt and pepper, stirring occasionally.

❺ Add the tuna, drained from the oil, and continue cooking for 5 minutes.

❻ Meanwhile bring the salted water to the boil. Add the spaghetti and cook for 7 minutes until al dente, stirring occasionally.

❼ Drain the spaghetti and toss with the sauce. Sprinkle the parsley over and toss well. Sauté for 1 minute and serve.

SPAGHETTI DEL GHIOTTONE

貪食的義大利麵

材料　　4～6人份	Ingredients Serves: 4-6
義大利麵 ·····················400公克	400 g spaghetti
酸豆·····························30公克	30 g capers
乾番茄························100公克	100 g sun-dried tomatoes
鯷魚····························30公克	30 g anchovy
大蒜切碎······················30公克	30 g garlic, finely chopped
辣椒切碎······················10公克	10 g fresh chili, finely chopped
荷蘭芹切碎···················50公克	50 g parsley, finely chopped
特級純橄欖油 ···············40公克	40 g extra-virgin olive oil
鹽、胡椒······················適量	salt and freshly ground pepper

作法

1 將乾番茄切成條狀。

2 橄欖油放入平底鍋內加熱，炒香大蒜數分鐘，放入鯷魚，用木杓拌炒至鯷魚融化。

3 接著加入荷蘭芹、辣椒、乾番茄，加入適量鹽、胡椒調味，並稍加攪拌。

4 另將麵條放入滾沸的鹽水中，煮10分鐘至麵條軟硬適中，須稍加攪動。

5 撈出麵條，與醬料翻炒均勻，即可盛盤上桌，並視各人喜好撒些帕梅善起司粉。

Preparation

1 Cut the sun-dried tomatoes into small julienne pieces.

2 Heat the olive oil in a frying pan. Add the garlic and sauté for few minutes. Add the anchovy and sauté until melted.

3 Add the parsley, chili and sun-dried tomatoes. Season with salt and pepper, stirring occasionally.

4 Bring the salted water to the boil. Add the spaghetti and cook for 10 minutes until al dente, stirring occasionally.

5 Drain the spaghetti and toss with the sauce. Serve immediately. Spread with Parmesan cheese to taste.

翡翠松子飯

材料	4～6人份
義式米	400公克
奶油	80公克
帕梅善起司粉	80公克
松子醬（見16頁）	150公克
紅蔥頭切碎	30公克
白酒	1杯
牛肉高湯	500cc
鹽	適量

Ingredients Serves: 4-6

400 g arborio rice
80 g butter
80 g freshly grated Parmesan cheese
150 g pesto sauce (see page 16)
30 g shallots, finely chopped
1 glass white wine
500 cc beef broth
salt

作法

❶ 鍋內倒入1/2奶油加熱，炒香紅蔥頭至變軟。

❷ 將米放入鍋內翻炒，倒入白酒，不斷攪拌至酒精完全揮發。

❸ 分次加入些許煮沸的牛肉高湯，不停攪拌至湯汁被米粒吸收。

❹ 重複（作法3）的步驟，至米飯煮到軟硬適中，約需16至18分鐘。

❺ 熄火前2分鐘，加入松子醬拌勻。

❻ 米飯煮好後，加入帕梅善起司和剩餘的奶油拌勻，即可趁熱上桌。

Preparation

❶ In a medium saucepan, heat half the butter and sauté the shallots until softened.

❷ Add the rice and stir well. Pour in the white wine. Stir well until the wine has evaporated.

❸ Add some boiled beef broth, stirring occasionally until all the liquid has been absorbed.

❹ Repeat the procedure until the rice is al dente, for 16-18 minutes.

❺ Add the pesto sauce 2 minutes before removing from the heat, stirring occasionally.

❻ When the rice is ready, add the Parmesan cheese and remaining butter. Stir well and serve hot.

魚子醬季諾多

材料　　4～6人份

季諾多麵	400公克
大蒜切碎	20公克
洋蔥切碎	30公克
鯷魚	20公克
綠花椰菜	800公克
荷蘭芹切碎	30公克
魚子醬	40公克
特級純橄欖油	40公克
鹽、胡椒	適量

Ingredients Serves: 4-6

400 g girandole
20 g garlic, finely chopped
30 g onions, finely chopped
20 g anchovy
800 g broccoli
30 g parsley, finely chopped
40 g caviar
40 g extra-virgin olive oil
salt and freshly ground pepper

作法

❶ 綠花椰菜洗淨,切成小朵。

❷ 將綠花椰菜放入滾沸的鹽水中,煮8分鐘,撈起瀝乾,將鹽水保留備用。

❸ 將橄欖油放入平底鍋內加熱,炒香大蒜和洋蔥至變軟,放入鯷魚,用木杓輕輕拌炒至鯷魚融化。

❹ 接著加入花椰菜,加適量鹽、胡椒,煮5分鐘並稍加攪拌。

❺ 另將麵條放入滾沸的鹽水中,煮8分鐘至麵條軟硬適中,煮時須稍加攪動,以免沾鍋。

❻ 撈出麵條,與醬料翻炒均勻,撒入荷蘭芹和魚子醬拌勻,續炒1分鐘,即可趁熱盛盤上桌。

Preparation

❶ Clean and wash the broccoli. Cut into florets.

❷ Bring the salted water to the boil. Add the broccoli and cook for 8 minutes. Reserve the salted water.

❸ Heat the olive oil in a frying pan. Add the garlic and onions, and cook until onions have softened. Add the anchovy and stir well until melted.

❹ Add the broccoli and sauté for 5 minutes. Season with salt and pepper, stirring occasionally.

❺ Bring the salted water to the boil. Add the girandole and cook for 8 minutes until al dente, stirring occasionally.

❻ Drain the girandole and toss with the sauce. Sprinkle over the parsley and caviar, and sauté for 1 minute. Serve immediately.

雙色麵疙瘩

材料　　4～6人份

麵疙瘩部分：
馬鈴薯·····························1公斤
中筋麵粉·····················400公克
橄欖油·····························1大匙
蛋·····································1個

醬料部分：
蘆筍去皮·····················800公克
帕瑪火腿切條···············200公克
番紅花·····························0.3公克
奶油·······························50公克
茱萸蔥切碎·····················25公克
鹽、胡椒·····························適量

Ingredients Serves: 4-6

For the gnocchi:
1 kg potatoes
400 g all-purpose flour
1 tbsp olive oil
1 egg
For the sauce:
800 g asparagus, trimmed and peeled
200 g Parma ham, cut in strips
0.3 g saffron
50 g butter
25 g chive, coarsely chopped
salt and freshly ground pepper

作法

❶ 馬鈴薯帶皮放入沸水中，煮約40分鐘，待冷卻後去皮，搗成泥。

❷ 將馬鈴薯泥、蛋、麵粉及橄欖油混合均勻，搓揉成柔軟光滑的麵團。

❸ 取一小塊麵團搓成手指粗細的長條，切成2公分長的小段。

❹ 用叉子或其他適當器具將麵疙瘩壓出花紋，捲成環狀，撒些麵粉，以免相黏。

❺ 將蘆筍放入煮沸的鹽水中，煮約6至7分鐘，撈起沖冷水使其冷卻，瀝乾水分，縱向剖半後，再切成3小段。

❻ 平底鍋內放入奶油加熱，加入帕瑪火腿炒2分鐘，再加蘆筍續煮3分鐘。

❼ 將番紅花與些許高湯調勻，加入蘆筍與火腿中，加適量鹽、胡椒調味，煮至湯汁收乾後熄火，煮時須稍加攪拌。

❽ 麵疙瘩放入滾沸的鹽水中，煮3至4分鐘並稍加攪動，直到麵疙瘩浮上水面。

❾ 將麵疙瘩撈出，與醬料翻炒1分鐘，撒入茱萸蔥，即可趁熱盛盤上桌。

Preparation

❶ Cook the potatoes in boiling water for 40 minutes. Drain, let cool, peel and mash the potatoes.

❷ Knead the mashed potatoes with the egg, flour and olive oil gently, until the mixture is well mixed and smooth.

❸ Take a small handful of dough, roll into strips and cut crosswise into 2cm long pieces.

❹ Press each piece of dough to make the curve with a fork or any proper utensil, and dip in flour. There should be an indentation on one side and imprint of fork prongs on the other side. Roll into rings and keep each piece separated on the floured board.

❺ Bring the salted water to the boil. Add the asparagus and cook for 6-7 minutes until al dente. Let cool and drain well. Cut the asparagus lengthwise into 3 pieces.

❻ Heat the butter in a frying pan. Add the Parma ham and sauté for 2 minutes. Add the asparagus and sauté for 3 minutes.

❼ Melt the saffron with some broth, pour it into the asparagus, and season with salt and pepper. When the broth has evaporated, remove from the heat, stirring occasionally.

❽ Bring the salted water to the boil. Add the gnocchi. Cover with a lid and cook for 3-4 minutes until they rise to the surface, stirring occasionally.

❾ Drain the gnocchi, toss with the sauce and sauté for 1 minute. Sprinkle with the chive and serve immediately.

焗海鮮麵捲

材料	4～6人份

義大利麵團（見14頁）⋯400公克
板魚排去皮 ⋯⋯⋯⋯⋯⋯600公克
蝦仁 ⋯⋯⋯⋯⋯⋯⋯⋯⋯200公克
鮮奶油醬（見18頁）⋯⋯200公克
芥末醬 ⋯⋯⋯⋯⋯⋯⋯⋯30公克
白酒 ⋯⋯⋯⋯⋯⋯⋯⋯⋯1/2杯
奶油 ⋯⋯⋯⋯⋯⋯⋯⋯⋯70公克
紅蔥頭切碎 ⋯⋯⋯⋯⋯⋯30公克
鹽 ⋯⋯⋯⋯⋯⋯⋯⋯⋯⋯適量

Ingredients Serves: 4-6

400 g pasta dough (see page 14)
600 g sole fish fillet, peeled
200 g shrimps, peeled
200 g Bechamel sauce (see page 18)
30 g mustard
1/2 glass white wine
70 g butter
30 g shallots, chopped
salt

作法

❶ 平底鍋內倒入1/2奶油加熱，放入紅蔥頭炒軟，加入魚肉和蝦仁續煮數分鐘，倒入白酒，煮到酒精完全揮發。

❷ 取出魚肉和蝦仁，剁碎。

❸ 將魚肉放入碗內，加入芥末醬、鮮奶油醬和鹽拌勻。

❹ 將麵團擀成薄片，切成8×12公分的麵皮數片。

❺ 將麵皮放入滾沸的鹽水中，煮4分鐘，每次勿煮超過7片，並須稍加攪動。煮好用漏杓撈起，放入冷水浸泡後瀝乾，一片片平放在略濕的布巾上，並以保鮮膜覆蓋，以免麵皮變乾。

❻ 每片麵皮鋪放2大匙的醬料，捲起成麵捲。

❼ 烤盤刷上奶油，放入麵捲，上層放幾塊奶油，放入預熱200℃（400℉）的烤箱中，烤25至30分鐘，直到表面金黃即可。

Preparation

❶ Heat half the butter in a frying pan. Sauté the shallots until softened. Add the sole fillets and shrimps, and sauté for few minutes. Add the wine and cook until it has evaporated.

❷ Remove the fish and shrimps from the pan and coarsely chop.

❸ Place the fish in a bowl with mustard and Bechamel. Season with salt and mix well.

❹ Roll out the pasta dough into a thin sheet, and cut into 8×12cm pieces with a pastry wheel.

❺ Bring the salted water to the boil. Add the pasta sheet and cook for 4 minutes, stirring occasionally. Do not cook more than 7 pieces at a time. Drain with a sieve and immerse in cold water. Drain again. Arrange the pasta sheet one by one on a tray, lined with damped towel and covered with plastic wrap.

❻ Spread each pasta sheet with 2 tablespoons of fish mixture and roll into cylinders.

❼ Butter the baking dish. Put the Canelloni in each baking pan. Spread over some pieces of butter. Heat the oven to 200℃(400℉) and bake the Canelloni for 25-30 minutes until nicely browned.

秋
Autumn

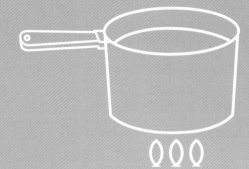

煮一鍋水傳遞濃濃暖意

用麵棍撫平起皺的思緒

撒一把調味料豐富人生的滋味

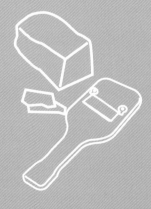

飲一杯好酒整理旅途中的心情

甜椒茄子麵

材料　　4～6人份

螺旋麵	400公克
茄子	350公克
罐頭番茄去皮、切碎	400公克
黃椒	2個
酸豆	30公克
黑橄欖	100公克
大蒜切碎	30公克
九層塔切碎	30公克
特級純橄欖油	40公克
鹽、胡椒	適量

Ingredients Serves: 4-6

400 g fusilli
350 g eggplants
400 g canned tomatoes, peeled and
coarsely chopped
2 yellow bell peppers
30 g capers
100 g black olives
30 g garlic, chopped
30 g basil, coarsely chopped
40 g extra-virgin olive oil
salt and freshly ground pepper

作法

❶ 將茄子洗淨後，切成小丁，在表面撒一層鹽，放置於濾網上1小時，使水分濾出。

❷ 黃椒以大火烤至外皮完全變黑。烘烤時須持續翻動。

❸ 待黃椒冷卻後，剝除外皮並縱向對切，去核去籽，切碎備用。

❹ 橄欖油倒入平底鍋內加熱，炒香大蒜，再放茄子煮數分鐘。

❺ 接著加入番茄、橄欖、酸豆、黃椒和九層塔，加適量鹽、胡椒調味，倒入些許的水，加蓋煮1小時直到蔬菜幾乎融化，煮時須持續攪拌。

❻ 另將螺旋麵放入滾沸的鹽水中，煮12分鐘至麵條軟硬適中，煮時須稍加攪動。撈出麵條，和蔬菜醬料翻炒均勻，即可趁熱上桌。

Preparation

① Wash the eggplants and cut into small cubes. Sprinkle with salt, put in a colander for 1 hour until the vegetation water goes out.

② Bake the bell peppers over high heat, turning them constantly until the skins are blackened completely.

③ Let cool the bell peppers, peel off the skin and cut lengthwise in half. Discard the cores and seeds. Coarsely chop.

④ Heat the olive oil in a frying pan. Add the garlic and sauté for few minutes. Add the eggplants and cook for few minutes.

⑤ Add the tomatoes, olives, capers, bell peppers and basil, seasoning with salt and pepper. Add some water, cover with a lid and simmer for 1 hour until the vegetables are almost melted, stirring occasionally.

⑥ Meanwhile bring the salted water to the boil. Add the fusilli and cook for 12 minutes until al dente, stirring occasionally. Drain the fusilli and toss with the sauce. Serve immediately.

迷迭香麵

材料　　4～6人份

筆型麵	400公克
培根	200公克
番茄	600公克
大蒜切碎	30公克
迷迭香	20公克
辣椒切碎	10公克
帕梅善起司粉	50公克
特級純橄欖油	40公克
鹽、胡椒	適量

Ingredients Serves: 4-6

400 g penne
200 g bacon
600 g ripe tomatoes
30 g garlic, chopped
20 g rosemary
10 g fresh chili, chopped
50 g freshly grated Parmesan cheese
40 g extra-virgin olive oil
salt and freshly ground pepper

作法

❶ 將培根切成條狀。

❷ 用刀在番茄末端劃出十字形，放入滾沸的熱水中煮約10秒鐘後撈起，剝皮、去籽、切碎。

❸ 橄欖油放入平底鍋內加熱，放入大蒜炒數分鐘，加入培根和迷迭香，拌炒5分鐘。

❹ 接著加入番茄和辣椒，並加適量鹽、胡椒調味，用小火煮20分鐘，煮時須稍加攪拌。

❺ 另將麵條放入滾沸的鹽水中，煮約12分鐘，至麵條軟硬適中，煮時須稍加攪動。

❻ 撈出麵條，與醬料翻炒1分鐘，趁熱盛盤，撒些帕梅善起司粉即可。

Preparation

❶ Cut the bacon into strips.

❷ Score an "X" on the base of each tomato. Immerse them in boiling water for 10 seconds and drain. Peel and cut crosswise in half. Squeeze the seeds and coarsely chop.

❸ Heat the olive oil in a frying pan. Add the garlic and sauté for few minutes. Add the bacon and rosemary and sauté for 5 minutes.

❹ Add the tomatoes and chili, season with salt and pepper, stirring occasionally, and simmer for 20 minutes.

❺ Meanwhile bring the salted water to the boil. Add the penne and cook for 12 minutes until al dente, stirring occasionally.

❻ Drain the penne, toss with the suace and sauté for 1 minute. Sprinkle the Parmesan cheese on the top and serve.

鄉村風味麵

材料　　4〜6人份

螺紋麵·····················400公克
松子醬（見16頁）········200公克
罐頭番茄去皮、切碎······400公克
大蒜切碎··················25公克
鼠尾草·····················20公克
迷迭香·····················20公克
俄力岡·····················20公克
松子·······················20公克
特級純橄欖油···············40公克
佩可瑞若起司粉············40公克
鹽、胡椒·····················適量

Ingredients Serves: 4-6

400 g gnocchetto
200 g pesto sauce (see page 16)
400 g canned tomatoes, peeled and chopped
25 g garlic, chopped
20 g sage
20 g rosemary
20 g oregano
20 g pine nuts
40 g extra-virgin olive oil
40 g freshly grated Pecorino cheese
salt and freshly ground pepper

作法

❶ 鼠尾草切成條狀。

❷ 橄欖油放入平底鍋內加熱，放入大蒜炒數分鐘，放入鼠尾草、迷迭香和俄力岡同煮。

❸ 加入番茄，撒些鹽、胡椒調味，持續拌炒約15分鐘。

❹ 接著加入松子和松子醬，攪拌均勻，用小火保溫。

❺ 另將麵條放入滾沸的鹽水中，煮8至10分鐘，至麵條軟硬適中，煮時須稍加攪動。

❻ 撈出麵條，與醬料快速翻炒1分鐘，撒入佩可瑞若起司粉，即可上桌。

Preparation

❶ Cut the sage into strips.

❷ Heat the olive oil in a frying pan. Add the garlic and sauté for few minutes. Add the sage, rosemary and oregano. Stir occasionally.

❸ Add the chopped tomatoes, salt and pepper. Stir occasionally and cook for 15 minutes.

❹ Add the pine nuts and pesto sauce. Stir well and keep the sauce warm over low heat.

❺ Meanwhile bring the salted water to the boil. Add the gnocchetto and cook for 8-10 minutes until al dente, stirring occasionally.

❻ Drain the gnocchetto, toss with the sauce quickly and sauté for 1 minute. Sprinkle with Pecorino cheese and serve.

媽媽風味麵

材料　4～6人份

筆形麵	400公克
牛肩肉切碎	350公克
雞肝切碎	100公克
紅蔥頭切碎	50公克
帕瑪火腿	200公克
蘋果去皮、切丁	1個
白酒	1杯
帕梅善起司粉	40公克
雞高湯	少許
奶油	20公克
特級純橄欖油	20公克
鹽、胡椒	適量

Ingredients Serves: 4-6

400 g penne
350 g beef chuck, coarsely ground
100 g chicken liver, chopped
50 g shallots, chopped
200 g Parma ham
1 apple, peeled and diced
1 glass white wine
40 g freshly grated Parmesan cheese
chicken broth
20 g butter
20 g extra-virgin olive oil
salt and freshly ground pepper

作法

❶ 將帕瑪火腿的肥肉部分切下並剁碎。

❷ 另將帕瑪火腿的瘦肉部分切成條狀。

❸ 將火腿肥肉、奶油和橄欖油放入平底鍋內加熱，放入蘋果、紅蔥頭，炒至紅蔥頭變軟。

❹ 加入牛肉、雞肝和火腿瘦肉，續煮2分鐘，倒入白酒，加適量鹽、胡椒調味，煮至酒精完全揮發。

❺ 接著一點一點地加入雞高湯，用小火慢煮25至30分鐘，須稍加攪拌。

❻ 另將麵條放入滾沸的鹽水中，煮8至10分鐘，至麵條軟硬適中，煮時須稍加攪動。

❼ 撈出麵條，與醬料翻炒均勻，撒入帕梅善起司粉再拌炒1分鐘，即可盛盤上桌。

Preparation

❶ Remove the fat from the Parma ham and chop finely.

❷ Cut the Parma ham into strips.

❸ Heat the Parma ham fat, butter and olive oil in a frying pan. Add the apples and shallots, and sauté until the shallots have softened. Stir occasionally.

❹ Add the beef, liver and Parma ham and sauté for 2 minutes. Add the wine, salt and pepper and stir until the wine has evaporated.

❺ Add little by little the chicken broth and cook over low heat for 25-30 minutes. Stir occasionally.

❻ Meanwhile bring the salted water to the boil. Add the penne and cook for 8-10 minutes until al dente, stirring occasionally.

❼ Drain the penne and toss with the sauce. Sprinkle with the Parmesan cheese, sauté for 1 minute and serve.

淡菜蟹肉麵

材料　　4～6人份

義式細麵 ·····················400公克
蟹肉·····················200公克
淡菜肉·····················200公克
大蒜切碎·····················30公克
辣椒切碎·····················10公克
番茄切碎·····················250公克
白酒·····················1杯
特級純橄欖油·····················30公克
荷蘭芹切碎·····················30公克
鹽、胡椒·····················適量

Ingredients Serves: 4-6

400 g linguini
200 g shelled crab
200 g shelled mussels
30 g garlic, chopped
10 g fresh chili, chopped
250 g ripe tomatoes, coarsely chopped
1 glass white wine
30 g extra-virgin olive oil
30 g parsley, chopped
salt and freshly ground pepper

作法

❶ 橄欖油放入平底鍋內加熱，放入大蒜炒2分鐘。

❷ 加入淡菜和蟹肉一同拌炒，倒入白酒，煮至酒精完全揮發。

❸ 接著加入辣椒、番茄，並加適量鹽、胡椒調味，煮10分鐘。

❹ 另將麵條放入滾沸的鹽水中，煮10分鐘至麵條軟硬適中，須稍加攪動。

❺ 撈出麵條，與醬料翻炒均勻，撒入荷蘭芹拌勻。

❻ 趁熱盛盤上桌，再擺些荷蘭芹裝飾即可。

Preparation

❶ In a large frying pan, heat the olive oil and sauté the garlic for 2 minutes.

❷ Add the mussels and crab, and toss with the garlic. Add the white wine and cook until it has evaporated.

❸ Add the chili and tomatoes, seasoning with salt and pepper, and cook for 10 minutes.

❹ Meanwhile bring the salted water to the boil. Add the linguini and cook for 10 minutes until al dente, stirring occasionally.

❺ Drain the linguini and toss with the sauce. Add the parsley and toss well.

❻ Serve immediately on a warm service dish, decorated with some parsley.

鮮蠔貝殼麵

材料　4~6人份

貝殼麵	400公克
新鮮香菇	300公克
檸檬皮屑	20公克
蠔肉	300公克
荷蘭芹切碎	30公克
特級純橄欖油	40公克
洋蔥切碎	20公克
大蒜切碎	20公克
檸檬汁	20公克
鹽、胡椒	適量

Ingredients Serves: 4-6

400 g conchiglie
300 g fresh mushrooms
20 g lemon rinds, grated
300 g shelled oysters
30 g parsley, chopped
40 g extra-virgin olive oil
20 g onions, chopped
20 g garlic, chopped
20 g lemon juice
salt and freshly ground pepper

作法

❶ 清除香菇的細砂，去梗，洗淨瀝乾，切薄片。

❷ 橄欖油放入平底鍋內加熱，放入洋蔥和大蒜炒軟後，加入香菇，續煮約5分鐘。

❸ 接著加入鮮蠔肉、檸檬汁和檸檬皮屑，煮10分鐘至水分收乾，加適量鹽、胡椒調味。

❹ 另將麵條放入滾沸的鹽水中，煮12分鐘至麵條軟硬適中，須稍加攪動。

❺ 撈出麵條，與醬料翻炒均勻，撒入荷蘭芹再拌炒1分鐘，即可趁熱上桌。

Preparation

❶ Clean the mushrooms of grit and sand. Cut off the stem ends. Wash quickly, drain and slice into thin pieces.

❷ Heat the olive oil in a frying pan. Add the onions and garlic, and sauté until the onions have softened. Add the mushrooms and sauté for 5 minutes.

❸ Add the oysters, lemon juice and lemon rinds. Cook for 10 minutes until all the water has evaporated. Season with salt and pepper.

❹ Meanwhile bring the salted water to the boil. Add the conchiglie and cook for 12 minutes until al dente, stirring occasionally.

❺ Drain the conchiglie and toss with the sauce. Sprinkle with the parsley and sauté for 1 minute. Serve hot.

鮪魚醬麵

材料　　4～6人份

筆形麵……………………400公克	
茄汁醬（見17頁）………300公克	
罐頭鮪魚瀝除油分………200公克	
莫札雷拉起司切丁………200公克	
黑橄欖切薄片……………50公克	
特級純橄欖油……………30公克	

Ingredients Serves: 4-6

400 g penne
300 g tomato sauce (see page 17)
200 g canned tuna, drained
200 g Mozzarella cheese, diced
50 g black olives, thinly sliced
30 g extra-virgin olive oil

作法

❶ 橄欖油放入平底鍋內加熱，放入茄汁醬煮沸，加入鮪魚和橄欖拌炒均勻。

❷ 另將麵條放入滾沸的鹽水中，煮約10分鐘，至麵條軟硬適中，煮時須稍加攪動。

❸ 撈出麵條，與醬料翻炒均勻。

❹ 加入莫札雷拉起司拌勻，即可上桌。

Preparation

❶ Heat the olive oil in a frying pan. Add the tomato sauce and bring to boil. Add the tuna and olives and stir well.

❷ Meanwhile bring the salted water to the boil. Add the penne and cook for 10 minutes until al dente, stirring occasionally.

❸ Drain the penne and toss with the sauce.

❹ Add the Mozzarella cheese, toss well and serve immediately.

a
b
c
d
e

RISOTTO CON RADICCHIO E ZAFFERANO

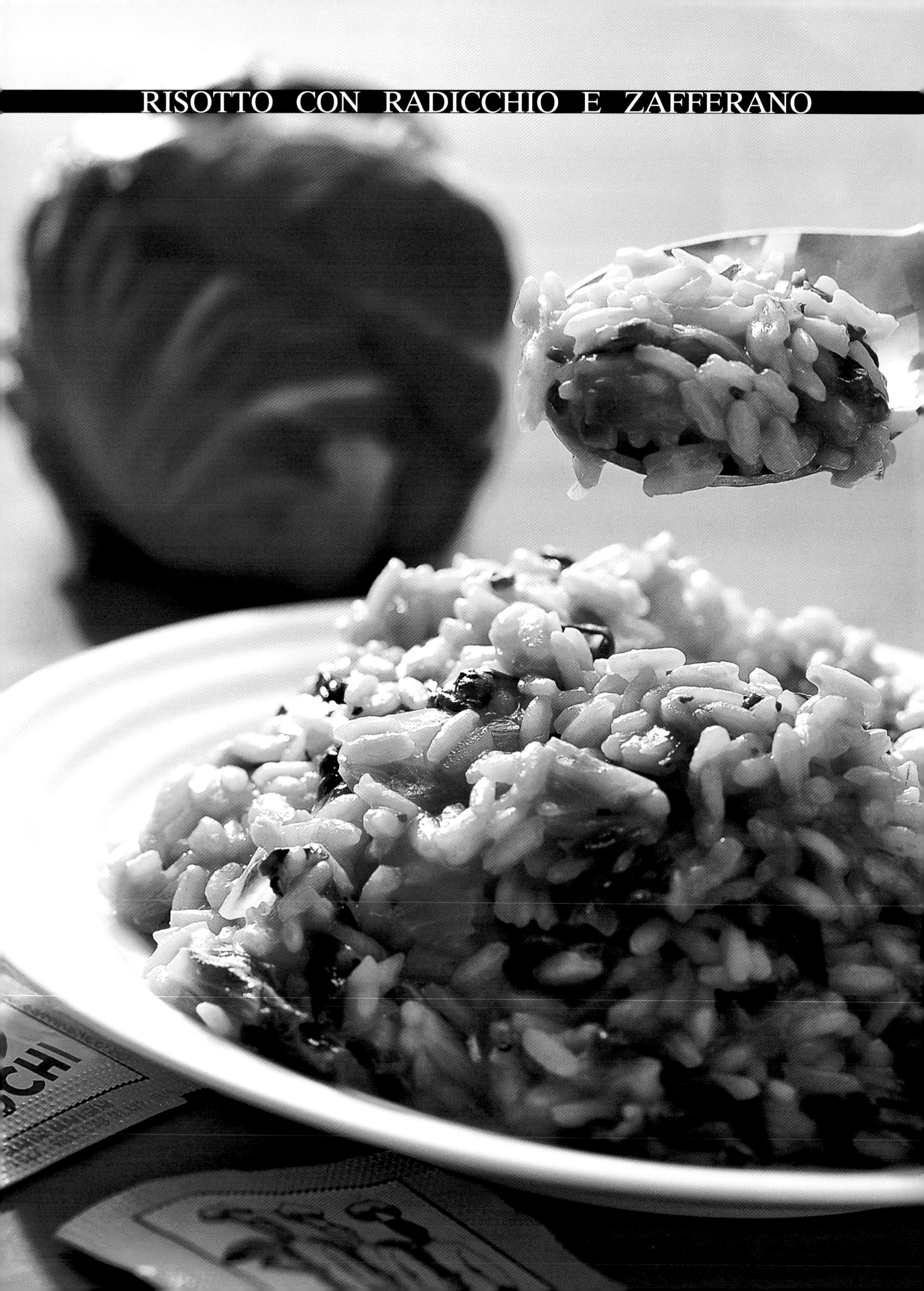

紫色米飯

材料	4～6人份	Ingredients Serves: 4-6
義式米	400公克	400 g arborio rice
紫萵苣	600公克	600 g radicchio
奶油	80公克	80 g butter
紅蔥頭切碎	30公克	30 g shallots, finely chopped
番紅花	0.3公克	0.3 g saffron
帕梅善起司粉	40公克	40 g freshly grated Parmesan cheese
白酒	1杯	1 glass white wine
牛肉高湯	500cc	500 cc beef broth
鹽、胡椒	適量	salt and freshly ground pepper

作法

❶ 切除紫萵苣的硬梗和過老的菜葉，切成細絲。

❷ 鍋內倒入1/2奶油加熱，放入紅蔥頭炒數分鐘，加入紫萵苣，倒入1/2白酒，撒鹽、胡椒調味，加蓋用小火煮15分鐘。若醬汁太乾，可加些許牛肉高湯調和。

❸ 將米放入鍋內翻炒均勻，並倒入剩餘的白酒，不斷攪拌至酒精完全揮發。

❹ 加入些許煮沸的牛肉高湯，持續攪拌至湯汁完全被米粒吸收。重複此一步驟，直到米飯軟硬適中，約需16至18分鐘。

❺ 用一小碗將番紅花與些許牛肉高湯拌勻，在米飯煮熟前2分鐘，將番紅花倒入拌勻。

❻ 熄火，加入剩餘的奶油和帕梅善起司粉，快速攪拌均勻，趁熱盛盤上桌。

Preparation

❶ Discard the tough stems and uncolored leaves of the radicchio. Cut in half and slice each part into thin strips.

❷ Heat half the butter in a heavy saucepan and sauté the shallots for few minutes. Add the radicchio, half the wine, salt and pepper. Cover with a lid and simmer for 15 minutes. If it is too dry, pour in some beef broth.

❸ Add the rice and stir well. Pour in the remaining wine and stir well until it has evaporated.

❹ Add the boiled beef broth, and stir occasionally until the liquid has been absorbed. Repeat the procedure until the rice is al dente, for 16-18 minutes.

❺ Melt the saffron with some beef broth in a small bowl. Pour it into the rice 2 minutes before ready.

❻ Remove the rice from the heat and pour in the remaining butter and Parmesan cheese. Stir well and quickly. Serve immediately.

托飛海鮮麵

材料　　4～6人份

托飛麵·····················400公克
淡菜肉·····················200公克
蛤蜊肉·····················200公克
蝦仁·······················200公克
松子醬（見16頁）·········180公克
大蒜切碎···················30公克
鮮奶油·····················100公克
帕梅善起司粉···············40公克
特級純橄欖油···············40公克
白酒·······················1/2杯
鹽、胡椒···················適量

Ingredients Serves: 4-6

400 g trofie
200 g shelled mussels
200 g shelled clams
200 g peeled shrimp
180 g pesto sauce (see page 16)
30 g garlic, chopped
100 g whipping cream
40 g freshly grated Parmesan cheese
40 g extra-virgin olive oil
1/2 glass white wine
salt and freshly ground pepper

作法

❶ 橄欖油放入平底鍋內加熱，放入大蒜炒數分鐘，再放入淡菜肉和蛤蜊肉，翻炒3分鐘。

❷ 加入蝦仁續炒數分鐘後，倒入白酒，煮到酒精完全揮發，撒些鹽、胡椒調味，煮時須稍加攪拌。

❸ 加入松子醬，與海鮮料拌炒均勻並煮沸，改小火。

❹ 另將麵條放入滾沸的鹽水中，煮12分鐘至麵條軟硬適中，須稍加攪動。

❺ 撈出麵條，與醬料翻炒均勻。

❻ 撒上帕梅善起司粉，拌炒1分鐘，即可盛盤上桌。

Preparation

❶ Heat the olive oil in a frying pan. Add the garlic and sauté for few minutes. Add the mussels and clams, and sauté for 3 minutes.

❷ Add the shrimps and sauté for few minutes. Pour in the white wine and allow to evaporate. Season with salt and pepper, stirring occasionally.

❸ Add the pesto sauce, toss with the seafood and bring to boil. Reduce the heat.

❹ Meanwhile bring the salted water to the boil. Add the trofie and cook for 12 minutes until al dente, stirring occasionally.

❺ Drain the trofie and toss with the sauce.

❻ Sprinkle with Parmesan cheese, sauté for 1 minute and serve.

托羅麵疙瘩

材料	4～6人份	Ingredients Serves: 4-6

麵疙瘩部分：

馬鈴薯 ·················· 1公斤
中筋麵粉 ············ 400公克
蛋 ·························· 1個
橄欖油 ················· 1大匙

醬料部分：

煙燻火腿 ············· 150公克
新鮮香菇切薄片 ···200公克
紅蔥頭切碎 ··········· 30公克
鮮奶油 ················· 100公克
白酒 ····················· 1/2杯
奶油 ····················· 30公克
帕梅善起司粉 ········· 40公克
鹽、胡椒 ················· 適量

For the gnocchi:
1 kg potatoes
400 g all-purpose flour
1 egg
1 tbsp olive oil
For the sauce:
150 g speck (smoked ham)
200 g fresh mushrooms, thinly sliced
30 g shallots, chopped
100 g whipping cream
1/2 glass white wine
30 g butter
40 g freshly grated Parmesan cheese
salt and freshly ground pepper

作法

❶ 馬鈴薯帶皮放入沸水中，煮約40分鐘，待冷卻後去皮，搗成泥。

❷ 將馬鈴薯泥、蛋、麵粉及橄欖油混合均勻，搓揉成柔軟光滑的麵團。

❸ 取一小塊麵團搓成手指粗細的長條，切成2公分長的小段。

❹ 用叉子或其他適當器具將麵疙瘩壓出花紋，捲成環狀，撒些麵粉，以免相黏。

❺ 奶油放入平底鍋內加熱，放入紅蔥頭炒軟，加入香菇炒至水分收乾，煮時須稍加攪拌。

❻ 加火腿，撒鹽、胡椒調味，倒入白酒，煮到酒精完全揮發，加入鮮奶油，用小火續煮，注意勿煮沸。

❼ 另將麵疙瘩放入滾沸的鹽水中，加蓋煮3至4分鐘，須稍加攪動，直到麵疙瘩浮上水面。

❽ 將麵疙瘩撈起瀝乾，與醬料拌炒約1分鐘，趁熱盛盤，撒些帕梅善起司粉即可。

Preparation

❶ Cook the potatoes in boiling water for 40 minutes. Drain, let cool, peel and mash the potatoes.

❷ Knead the mashed potatoes with the egg, flour and olive oil gently, until the mixture is well mixed.

❸ Take a handful of the dough and roll into strips. Cut crosswise into 2cm long pieces.

❹ Press each piece of dough to make the curve with a fork or any proper utensil, and dip in flour. There should be an indentation on one side and imprint of fork prongs on the other side. Roll into rings and keep each piece separated on the floured board.

❺ Heat the butter in a frying pan. Add the shallots and sauté until soft. Add the mushrooms and cook until all the vegetation water has evaporated. Stir occasionally.

❻ Add the speck, salt, and pepper. Stir until the wine has evaporated. Add the cream and cook over low heat. Do not allow the sauce to boil.

❼ Meanwhile bring the salted water to the boil. Add the gnocchi, cover with a lid and cook for 3-4 minutes until they rise to the surface, stirring occasionally.

❽ Drain the gnocchi, toss with the sauce and sauté for 1 minute. Sprinkle with the Parmesan cheese and serve immediately.

烤菠菜麵捲

a

b

c

d

e

f

g

h

材料	4～6人份

菠菜煮熟、瀝乾、切碎⋯100公克
瑞柯達起司⋯⋯⋯⋯⋯250公克
帕梅善起司粉⋯⋯⋯⋯160公克
蛋⋯⋯⋯⋯⋯⋯⋯⋯⋯⋯1個
豆蔻粉⋯⋯⋯⋯⋯⋯⋯⋯適量
義大利麵團（見14頁）⋯300公克
奶油⋯⋯⋯⋯⋯⋯⋯⋯⋯50公克
鹽、胡椒⋯⋯⋯⋯⋯⋯⋯適量

Ingredients Serves: 4-6

100 g boiled spinach, drained and
finely chopped
250 g Ricotta cheese
160 g freshly grated Parmesan cheese
1 egg
freshly grated nutmeg
300 g pasta dough (see page 14)
50 g butter
salt and freshly ground pepper

作法

❶ 製作餡料：將瑞柯達起司、蛋、100公克帕梅善起司粉、菠菜、豆蔻粉、鹽和胡椒拌勻。

❷ 將麵團擀成薄片，切成8公分×12公分的麵皮數片。

❸ 將麵皮放入滾沸的鹽水中，煮4分鐘，每次勿煮超過7片，並須稍加攪動。煮好用漏杓撈起，入冷水浸泡後瀝乾，一片片平放在略濕的布巾上，以保鮮膜覆蓋，以免麵皮變乾。

❹ 每片麵皮放上2大匙的菠菜醬料，捲起成麵捲。

❺ 將每個麵捲各切成3段。

❻ 烤盤刷上奶油，放入約9段麵捲，表面放幾塊奶油、撒剩餘的帕梅善起司粉，放入預熱200℃（400℉）的烤箱中，烤約25分鐘，至表面金黃即可。

Preparation

❶ Mix the Ricotta cheese, eggs, 100 g of Parmesan cheese, chopped spinach, nutmeg, salt and pepper in a bowl.

❷ Roll out the dough into a thin sheet and cut into 8×12㎝ pieces.

❸ Bring the salted water to the boil. Add the pasta pieces and cook for 4 minutes, stirring occasionally. Do not cook more than 7 pieces at a time. Drain with a sieve. Put in cold water, then drain again. Arrange the pasta sheet one by one on trays, lined with dampened towel and covered with plastic wrap.

❹ Spread each pasta piece with 2 tablespoons of Ricotta mixture and roll into cylinders.

❺ Cut each cylinder into 3 pieces.

❻ Butter the individual baking dish. Put 9 cylinders in each dish and spread over some pieces of remaining butter and the remaining Parmesan cheese. Heat the oven to 200℃(400℉) and bake the Panzerotti for 25 minutes until nicely browned.

RAVIOLI DI FUNGHI NEL LORO SUGO

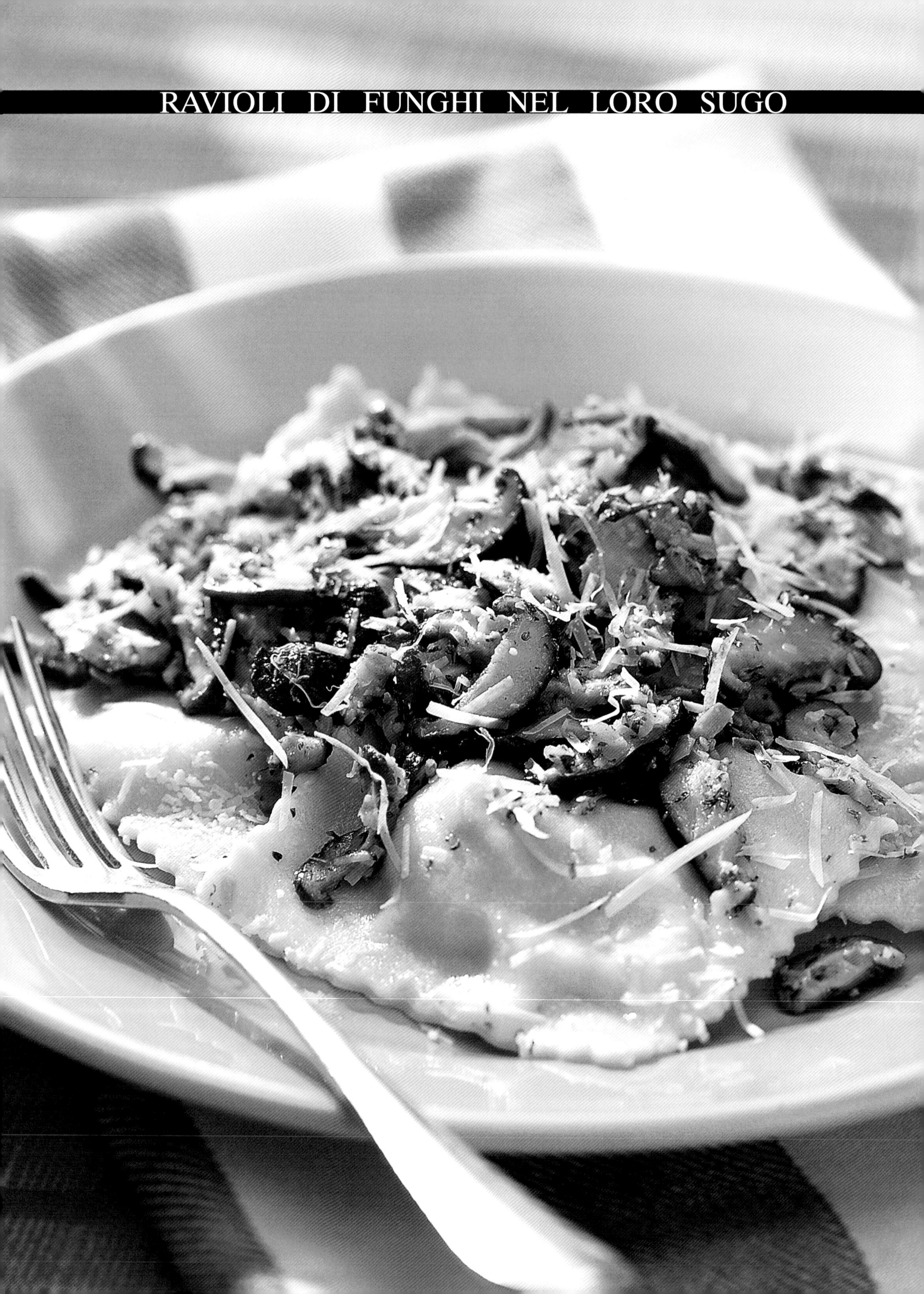

鮮香菇餃

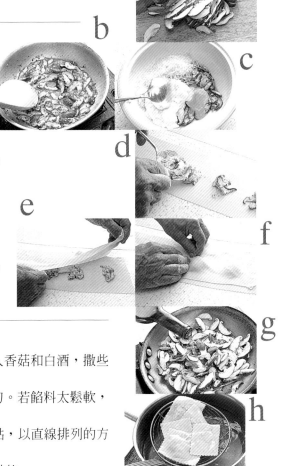

材料　　4～6人份

Ingredients Serves: 4-6

義大利麵團（見14頁）⋯400公克	400 g pasta dough (see page 14)
餡料部分：	For the stuffing:
新鮮香菇 ⋯⋯⋯⋯⋯⋯800公克	800 g fresh mushrooms
大蒜切碎⋯⋯⋯⋯⋯⋯⋯30公克	30 g garlic, chopped
紅蔥頭切碎 ⋯⋯⋯⋯⋯30公克	30 g shallots, chopped
荷蘭芹切碎 ⋯⋯⋯⋯⋯30公克	30 g parsley, chopped
白酒⋯⋯⋯⋯⋯⋯⋯⋯⋯1杯	1 glass white wine
蛋 ⋯⋯⋯⋯⋯⋯⋯⋯⋯⋯1個	1 egg
奶油⋯⋯⋯⋯⋯⋯⋯⋯40公克	40 g butter
帕梅善起司粉 ⋯⋯⋯⋯⋯80公克	80 g freshly grated Parmesan cheese
鮮奶油醬（見18頁）⋯⋯200公克	200 g Bechamel sauce (see page 18)
醬料部分：	For the sauce:
新鮮香菇 ⋯⋯⋯⋯⋯⋯400公克	400 g fresh mushrooms
大蒜切碎⋯⋯⋯⋯⋯⋯⋯30公克	30 g garlic, chopped
奶油⋯⋯⋯⋯⋯⋯⋯⋯40公克	40 g butter
白酒 ⋯⋯⋯⋯⋯⋯⋯⋯1/2杯	1/2 glass white wine
荷蘭芹切碎 ⋯⋯⋯⋯⋯30公克	30 g parsley, chopped
帕梅善起司粉 ⋯⋯⋯⋯⋯40公克	40 g freshly grated Parmesan cheese
鹽、胡椒⋯⋯⋯⋯⋯⋯⋯適量	salt and freshly ground pepper

作法

❶ 將所有香菇清除沙粒，切除梗部，洗淨後瀝乾，切薄片備用。

❷ 奶油放入平底鍋內加熱，放入大蒜、紅蔥頭和荷蘭芹炒軟，加入香菇和白酒，撒些鹽、胡椒調味，煮到酒精完全揮發。

❸ 將炒好的香菇放入碗內，加入帕梅善起司粉、鮮奶油醬和蛋拌勻。若餡料太鬆軟，可加些麵包粉拌勻。

❹ 取一塊麵團擀成6至7公分寬的麵皮，在四周邊緣刷些水以便沾黏，以直線排列的方式，每隔3公分放一大匙餡料，再用另一張麵皮覆蓋。

❺ 壓出空氣，用滾輪刀將麵皮切成3公分見方的正方形，四周壓緊黏住。

❻ 製作醬料：奶油放入平底鍋加熱，放入大蒜及荷蘭芹炒3至4分鐘，再加香菇拌炒，倒入白酒，撒鹽、胡椒調味，煮到酒精完全揮發。煮時須稍加攪拌。

❼ 另將香菇餃放入滾沸的鹽水中，煮6至8分鐘，煮時須稍加攪動，待熟，用漏杓撈起。

❽ 將香菇餃與醬料翻炒均勻，撒入帕梅善起司粉拌炒1分鐘，即可趁熱盛盤上桌。

Preparation

❶ Clean all the mushrooms of grit and sand. Cut off the stem ends and wash. Drain and cut in thin slices.

❷ Heat the butter in a frying pan. Add the garlic, shallots and parsley. Sauté until the shallots are softened. Stir occasionally. Add the mushrooms, wine, salt and pepper. Cook until the wine and vegetation water have evaporated.

❸ Mix the mushrooms in a bowl with Parmesan cheese, Bechamel and egg. If the stuffing is too soft, add some spoons of bread crumbs.

❹ Roll out a strip of pasta into 6-7㎝ thin sheet. Brush the edges of each sheet with water for sticking. Place a tablespoon of mushroom mixture at 3㎝ intervals along the pasta. Cover with the second pasta sheet.

❺ Press out the air pockets. With a pastry wheel, cut the pasta into 3×3㎝ squares. Press around each parcel to seal the edges.

❻ For the sauce: Heat the butter in a saucepan. Add the garlic and parsley and sauté for 3-4 minutes. Add the sliced mushrooms and toss with the garlic and parsley. Add the wine, salt and pepper, and cook until all the wine and vegetation water have evaporated. Stir occasionally.

❼ Meanwhile bring the salted water to the boil. Add the ravioli and cook until al dente for 6-8 minutes. Stir occasionally. Drain with a sieve.

❽ Toss the ravioli with the sauce. Sprinkle over Parmesan cheese and sauté for 1 minute. Serve immediately.

冬
Winter

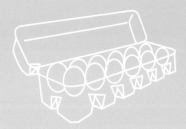

如果人生像義大利麵

軟硬適中才會愈嚼愈有味

如果愛情也像義大利麵

但願它永遠雋永好滋味

肉醬波紋麵

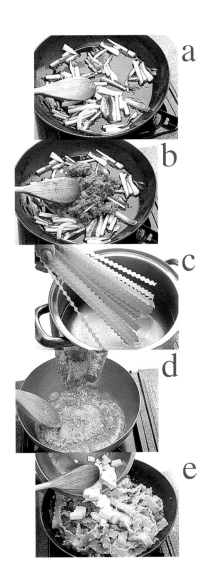

材料　　4～6人份	
波紋寬麵	400公克
節瓜切條	500公克
義式肉醬（見19頁）	400公克
莫札雷拉起司切丁	100公克
帕梅善起司粉	40公克
特級純橄欖油	40公克
麵包粉	40公克
奶油	40公克
鹽、胡椒	適量

Ingredients Serves: 4-6

400 g reginette
500 g zucchini, cut in strips
400 g Bolognese sauce (see page 19)
100 g Mozzarella cheese, diced
40 g freshly grated Parmesan cheese
40 g extra-virgin olive oil
40 g bread crumbs
40 g butter
salt and freshly ground pepper

作法

❶ 將橄欖油放入平底鍋內加熱，放入節瓜炒約6分鐘，直到呈金黃色。

❷ 加入肉醬，續煮數分鐘，撒些鹽、胡椒調味。如果醬汁太乾，可加些熱水調和。

❸ 另將麵條放入滾沸的鹽水中，煮12分鐘至麵條軟硬適中，須稍加攪拌。

❹ 用一小鍋加熱奶油，放入麵包粉，炒約2分鐘。

❺ 撈出麵條，與肉醬翻炒均勻，加入莫札雷拉起司、帕梅善起司及炒過的麵包粉拌勻，即可趁熱上桌。

Preparation

❶ Heat the olive oil in a frying pan. Add the zucchini and sauté for 6 minutes, until golden browned.

❷ Add the Bolognese sauce and sauté for few minutes. Adjust the salt and pepper. If the sauce is too dry, add some hot water.

❸ Meanwhile bring the salted water to the boil. Cook the reginette for 12 minutes until al dente, stirring occasionally.

❹ In a small saucepan, heat the butter, add the bread crumbs and sauté for 2 minutes.

❺ Drain the reginette and toss with the sauce. Add the Mozzarella, Parmesan cheese and bread crumbs. Toss well and serve.

FARFALLE ALLA PIEMONTESE

皮耶默皮蝶麵

材料　　4～6人份

蝴蝶麵················400公克
黃椒切丁 ················200公克
紅椒切丁 ················200公克
青椒切丁 ················200公克
番茄去籽、切丁 ·········300公克
火腿切細絲 ···············200公克
鮮奶油··················150公克
帕梅善起司粉 ·············40公克
特級純橄欖油 ·············40公克
鹽、胡椒··················適量

Ingredients Serves: 4-6

400 g farfalle
200 g yellow bell pepper, diced
200 g red bell pepper, diced
200 g green bell pepper, diced
300 g ripe tomatoes, seeded and diced
200 g ham, cut in juliennes
150 g whipping cream
40 g freshly grated Parmesan cheese
40 g extra-virgin olive oil
salt and freshly ground pepper

作法

❶ 橄欖油放入平底鍋內加熱，放入甜椒及番茄，煮15分鐘，撒些鹽、胡椒調味，煮時須稍加攪拌。

❷ 將煮好的醬料放入調理機內，加入鮮奶油攪拌均勻，撒適量鹽、胡椒調味，保溫備用。

❸ 另將麵條放入滾沸的鹽水中，煮12分鐘至麵條軟硬適中，須稍加攪動。

❹ 將醬料倒回鍋內煮沸。撈出麵條，與醬料翻炒拌勻。

❺ 撒入帕梅善起司粉和火腿，拌炒1分鐘，即可盛盤上桌。

Preparation

❶ Heat the olive oil in a frying pan. Add the bell peppers and tomatoes and cook for 15 minutes. Season with salt and pepper, stirring occasionally.

❷ Pour the sauce in a food processor, add the whipping cream and mix well. Adjust the salt and pepper. Keep warm.

❸ Bring the salted water to the boil. Add the farfalle and cook for 12 minutes until al dente. Stir occasionally.

❹ Bring the sauce to the boil. Drain the farfalle and toss with the sauce.

❺ Sprinkle over the Parmesan cheese and ham, and toss well. Sauté for 1 minute and serve.

鮮菇筆形麵

材料　　4～6人份	Ingredients Serves: 4-6
中型筆形麵 ……………400公克	400 g mezze penne
新鮮香菇 ……………250公克	250 g fresh mushrooms
瑞柯達起司 …………150公克	150 g Ricotta cheese
瑞士起司粉 …………150公克	150 g freshly grated Swiss cheese
帕梅善起司粉 ………50公克	50 g freshly grated Parmesan cheese
鼠尾草切碎 …………20公克	20 g sage, chopped
九層塔切碎 …………20公克	20 g basil, chopped
乾燥俄力岡 …………10公克	10 g dry oregano
荷蘭芹切碎 …………20公克	20 g parsley, chopped
奶油……………………60公克	60 g butter
鹽、胡椒……………適量	salt and freshly ground pepper

作法

❶ 將香菇清除沙粒，去梗，洗淨後瀝乾，切薄片。

❷ 碗內放入瑞士起司、瑞柯達起司、荷蘭芹、鼠尾草、九層塔和俄力岡拌勻。

❸ 平底鍋內倒入1/2奶油加熱，放入香菇炒5分鐘，撒入適量鹽、胡椒調味，煮時須稍加攪拌。

❹ 另將麵條放入滾沸的鹽水中，煮7分鐘至麵條軟硬適中，撈出麵條，與香菇拌勻，再加入（作法2）的起司料混合均勻。

❺ 將剩餘的奶油融化，刷在耐熱紙上。將耐熱紙鋪在烤盤內，再均勻擺入拌好的麵，並撒些帕梅善起司粉。

❻ 將耐熱紙包裹好，放入預熱至200℃（400℉）的烤箱內，烤5至6分鐘。

❼ 上桌前將耐熱紙剪開，即可盛盤食用。

Preparation

❶ Clean the mushrooms of grit and sand. Cut off the stems and wash. Drain and slice into thin pieces.

❷ In a bowl, mix well the Swiss cheese, Ricotta cheese, parsley, sage, basil and oregano.

❸ Heat half the butter in a frying pan. Add the mushrooms and sauté for 5 minutes. Season with salt and pepper, stirring occasionally.

❹ Meanwhile bring the salted water to the boil. Add the mezze penne and cook for 7 minutes until al dente. Drain and toss with the mushrooms. Add the cheese mixture and mix well.

❺ Melt the butter in a small saucepan. Brush a piece of oven paper with butter and put it on a baking dish. Spoon the pasta on the oven paper. Sprinkle the top with the Parmesan cheese.

❻ Wrap the pasta with the excess paper. Heat the oven to 200℃(400℉) and bake the pasta for 5-6 minutes.

❼ Unwrap the pasta and serve.

Colored Conchiglie

彩繪貝殼麵

材料　4~6人份

貝殼麵	400公克
芹菜切碎	100公克
胡蘿蔔切碎	100公克
洋蔥切碎	50公克
大蒜切碎	30公克
番茄	4個
荷蘭芹切碎	40公克
乾燥俄力岡	20公克
番茄糊	50公克
帕梅善起司粉	50公克
特級純橄欖油	40公克
鹽、胡椒	適量

Ingredients Serves: 4-6

400 g conchiglie
100 g celery, coarsely chopped
100 g carrots, coarsely chopped
50 g onions, chopped
30 g garlic, chopped
4 ripe tomatoes
40 g parsley, chopped
20 g dry oregano
50 g tomato paste
50 g freshly grated Parmesan cheese
40 g extra-virgin olive oil
salt and freshly ground pepper

作法

❶ 用刀在番茄末端劃出十字形，放入滾沸的熱水中燙數秒鐘，撈起剝皮、去籽，切碎。

❷ 橄欖油放入平底鍋內加熱，放入胡蘿蔔及芹菜炒5分鐘後，加入洋蔥和大蒜，用小火煮到洋蔥變軟且呈金黃色。

❸ 加入番茄、荷蘭芹和番茄糊，若醬汁太乾，可加入1/2杯熱水調和。再加適量鹽、胡椒調味，加蓋用小火燉煮25分鐘，並須稍加攪拌，煮好後加入俄力岡。

❹ 另將麵條放入滾沸的鹽水中，煮10至12分鐘，至麵條軟硬適中，煮時須稍加攪動，以免沾鍋。

❺ 撈出麵條，與醬料翻炒均勻，撒些帕梅善起司粉即可。

Preparation

❶ Score an "X" on the base of each tomato. Immerse them in boiling water for few seconds and drain. Peel and cut crosswise in half. Squeeze the seeds and finely chop.

❷ Heat the olive oil in a large frying pan. Add the carrots and celery and sauté for 5 minutes. Add the onions and garlic and cook over low heat, until the onions are softened and browned.

❸ Add the tomatoes, parsley and tomato paste. If the sauce is too dry, add 1/2 cup of hot water. Season with salt and pepper, cover with a lid and simmer for 25 minutes, stirring occasionally. When the sauce is ready, add the oregano.

❹ Meanwhile bring the salted water to the boil. Add the conchiglie and cook for 10-12 minutes until al dente, stirring occasionally.

❺ Drain the conchiglie and toss with the sauce. Sprinkle with the Parmesan cheese and serve.

SPAGHETTI AL CARTOCCIO CON VONGOLE

鑲裹義大利麵

材料　　4～6人份

義大利麵 ···················400公克
蛤蜊肉··················300公克
大蒜切片·················30公克
鯷魚··················30公克
茄汁醬（見17頁）·······250公克
白酒····················1杯
荷蘭芹切碎··············30公克
九層塔切碎··············30公克
特級純橄欖油············60公克
鹽、胡椒················適量

Ingredients Serves: 4-6

400 g spaghetti
300 g shelled clams
30 g garlic, sliced
30 g anchovy
250 g tomato sauce (see page 17)
1 glass white wine
30 g parsley, chopped
30 g basil, chopped
60 g extra-virgin olive oil
salt and freshly ground pepper

作法

❶ 平底鍋內加熱1/2橄欖油，炒香大蒜至金黃，約需2分鐘，再加入鯷魚，用木杓翻炒至融化。

❷ 放入蛤蜊肉，續煮1分鐘，倒入白酒煮到酒精完全揮發，再加入茄汁醬。

❸ 撒入適量鹽、胡椒調味，加入3/4的荷蘭芹和九層塔，用中火煮10分鐘。

❹ 另將麵條放入滾沸的鹽水中，煮時須稍加攪動，煮約8分鐘至麵條軟硬適中，撈出麵條，與醬料拌炒均勻。

❺ 烤盤內鋪一層錫箔紙，將炒好的麵均勻鋪放在錫箔紙上，淋入剩餘的橄欖油，將錫箔紙包裹好。

❻ 放入預熱至200℃（400℉）的烤箱內，烤10分鐘，取出，打開錫箔紙，撒入剩餘的荷蘭芹和九層塔，即可盛盤上桌。

Preparation

❶ In a medium frying pan, heat half the olive oil, add the garlic and sauté for 2 minutes, until the garlic turns brown. Add the anchovy, stir with a wooden spoon and cook until melted.

❷ Add the clams and sauté for 1 minute. Add the white wine and cook until it has evaporated. Add the tomato sauce and stir well.

❸ Season with salt and pepper, add 3/4 of the parsley and 3/4 of the basil, and cook for 10 minutes over medium heat.

❹ Meanwhile bring the salted water to the boil. Add the spaghetti and cook for 8 minutes until al dente, stirring occasionally. Drain the spaghetti and toss with the sauce.

❺ Spread the aluminum foil in a baking dish. Arrange the pasta over the foil and spread over the remaining olive oil. Wrap the pasta with the foil.

❻ Heat the oven of 200℃(400℉) and bake the pasta for 10 minutes. Unrap the pasta, spread over the remaining basil and parsley and serve.

LINGUINI BELLA MARIA

瑪莉亞海鮮麵

材料　　4～6人份	Ingredients Serves: 4-6
義式細麵 ·················400公克	400 g linguini
陸可拉（或九層塔）切絲100公克	100 g rucola (or basil), cut in strips
蝦仁·····················100公克	100 g peeled shrimps
蛤蜊肉···················100公克	100 g shelled clams
淡菜肉···················150公克	150 g shelled mussels
透抽切圈狀 ···············150公克	150 g squids, cut in rings
大蒜切碎·················20公克	20 g garlic, chopped
辣椒切碎·················10公克	10 g fresh chili, chopped
特級純橄欖油 ···········40公克	40 g extra-virgin olive oil
白蘭地···················30公克	30 g cognac
荷蘭芹切碎 ···············30公克	30 g parsley, chopped
鹽、胡椒·················適量	salt and freshly ground pepper

作法

❶ 橄欖油放入平底鍋內加熱，放入大蒜、辣椒和荷蘭芹，炒數分鐘，加入透抽，續炒5分鐘。

❷ 加入淡菜、蛤蜊肉和蝦仁同炒，並撒適量鹽、胡椒調味，煮到湯汁收乾。

❸ 接著倒入白蘭地，煮到酒精完全揮發。

❹ 另將麵條放入滾沸的鹽水中，煮8分鐘至麵條軟硬適中，煮時須稍加攪動，以免沾鍋。

❺ 撈出麵條，與醬料翻炒均勻，撒入陸可拉拌炒1分鐘，即可趁熱上桌。

Preparation

❶ Heat the olive oil in a frying pan. Add the garlic, chili and parsley and sauté for few minutes. Add the squids and cook for 5 minutes.

❷ Add the mussels, clams and shrimps. Season with salt and pepper, stirring occasionally. Sauté until all the water has evaporated.

❸ Add the cognac and let to evaporate.

❹ Bring the salted water to the boil. Add the linguini and cook for 8 minutes until al dente, stirring occasionally.

❺ Drain the linguini and toss with the sauce. Sprinkle over the rucola and sauté for 1 minute. Serve immediately.

Risotto with Bacon and Leeks

培根蒜苗飯

材料	4～6人份
義式米	400公克
奶油	80公克
培根切條	200公克
蒜苗切小段	500公克
白酒	1杯
紅蔥頭切碎	30公克
牛肉高湯	500cc
帕梅善起司粉	60公克
鹽、胡椒	適量

Ingredients Serves: 4-6

400 g arborio rice
80 g butter
200 g bacon, cut in strips
500 g leeks, cut short
1 glass white wine
30 g shallots, finely chopped
500 cc beef broth
60 g freshly grated Parmesan cheese
salt and freshly ground pepper

作法

❶ 平底鍋內倒入1/3奶油加熱，放入蒜苗炒軟，約需8分鐘，倒入1/2白酒，不斷攪拌至酒精完全揮發。

❷ 加入培根同炒5分鐘，撒鹽、胡椒調味，熄火備用。

❸ 另起鍋，加熱1/3奶油，放入紅蔥頭炒軟，煮約2至3分鐘。

❹ 接著加入米翻炒，並倒入剩餘的白酒，不斷攪拌至酒精完全揮發。

❺ 分次加入些許煮沸的牛肉高湯，持續攪拌至湯汁被米粒吸收。重複前述步驟，直到米飯軟硬適中，約需16至18分鐘。

❻ 熄火前5分鐘，放入（作法2）的蒜苗培根攪拌均勻。

❼ 最後加入剩餘的奶油和帕梅善起司粉拌勻，即可趁熱上桌。

Preparation

❶ In a frying pan, heat 1/3 of the butter and sauté the leeks for 8 minutes. Add half the white wine and stir occasionally until the wine has evaporated.

❷ Add the bacon and cook for 5 minutes, stirring constantly. Season with salt and pepper.

❸ In a heavy saucepan, heat 1/3 of the butter. Sauté the shallots until softened for 2-3 minutes.

❹ Add the rice and stir well. Pour in the remaining white wine. Stir well until the wine has evaporated.

❺ Add the boiled beef broth, stirring constantly until the liquid has been absorbed. Repeat the procedure until the rice is al dente for 16-18 minutes.

❻ 5 minutes before the rice is ready, add the leeks and bacon.

❼ Remove from the heat, and add the remaining butter and Parmesan cheese, stirring well and quickly. Serve immediately.

火腿紅花麵

材料	4～6人份

Ingredients Serves: 4-6

義式寬麵 ················400公克	400 g fetuccini
鮮奶油················200公克	200 g whipping cream
番紅花················0.3公克	0.3 g saffron
煙燻火腿 ···············200公克	200 g speck (smoked ham)
帕梅善起司粉 ···········40公克	40 g freshly grated Parmesan cheese
奶油··················30公克	30 g butter
豆蔻粉················適量	freshly grated nutmeg
鹽、胡椒···············適量	salt and freshly ground pepper

a
b
c
d
e
f

作法

1. 火腿切丁。

2. 奶油放入平底鍋內加熱，放入火腿炒數分鐘，注意勿炒太久，以免過鹹。倒入鮮奶油，撒適量鹽、胡椒和豆蔻粉調味，煮時須稍加攪拌。

3. 醬料煮沸時，加入番紅花拌炒均勻，改小火。

4. 另將麵條放入滾沸的鹽水中，煮12分鐘至麵條軟硬適中，須稍加攪動。

5. 撈出麵條，與醬料翻炒均勻。

6. 撒入帕梅善起司粉，拌炒1分鐘，即可趁熱盛盤上桌。

Preparation

1. Cut the speck into cubes.

2. Heat the butter in a frying pan. Add the speck and sauté for few minutes. Do not overcook, or it would become too salty. Add the cream, salt, pepper and nutmeg, stirring occasionally.

3. When the sauce starts to boil, add the saffron, stirring occasionally. Reduce the heat.

4. Bring the salted water to the boil. Add the fetuccini and cook for 12 minutes until al dente, stirring occasionally.

5. Drain the fetuccini and toss with the sauce.

6. Sprinkle with Parmesan cheese, sauté for 1 minute and serve.

山瑞莫雞冠麵

材料　4～6人份

雞冠麵……………………400公克
番茄去皮、去籽、切碎…300公克
黑橄欖去籽………………200公克
松子醬（見16頁）………150公克
特級純橄欖油……………40公克
佩可瑞若起司粉…………40公克
鹽、胡椒…………………適量

Ingredients Serves: 4-6

400 g Creste di Gallo
300 g ripe tomatoes, peeled, seeded and chopped
200 g black olives, seeded
150 g pesto sauce(see page 16)
40 g extra-virgin olive oil
40 g freshly grated Pecorino cheese
salt and freshly ground pepper

作法

❶ 將番茄、橄欖、松子醬、鹽、胡椒和橄欖油，拌勻備用。

❷ 另將麵條放入滾沸的鹽水中，煮8分鐘至麵條軟硬適中，須稍加攪動。

❸ 撈出麵條，與醬料翻炒均勻。

❹ 撒入佩可瑞若起司拌勻，即可盛盤上桌。

Preparation

❶ Put the tomatoes, olives, pesto, salt, pepper and olive oil in a bowl. Mix well.

❷ Bring the salted water to the boil. Add the Creste di Gallo and cook for 8 minutes until al dente, stirring occasionally.

❸ Drain the Creste di Gallo and toss quickly with the sauce.

❹ Sprinkle and toss with the Pecorino cheese. Serve immediately.

焗松子醬千層麵

材料　　4～6人份

菠菜麵團（見15頁）……600公克
鮮奶油醬（見18頁）……300公克
松子醬（見16頁）………250公克
佩可瑞若起司粉…………100公克
奶油………………………50公克

Ingredients Serves: 4-6

600 g spinach pasta dough (see page 15)
300 g Bechamel (see page 18)
250 g pesto sauce (see page 16)
100 g freshly grated Pecorino cheese
50 g butter

a
b
c
d
e
f

作法

❶ 將麵團擀成薄片，切成8×12公分的麵皮數片。

❷ 將麵皮放入滾沸的鹽水中煮4分鐘，每次勿煮超過7片，並須稍加攪動。煮好後用漏杓撈起，入冷水浸泡後瀝乾，一片片平放在略濕的布巾上，以保鮮膜覆蓋，以免麵皮變乾。

❸ 將松子醬、鮮奶油醬和1/2佩可瑞若起司拌勻。

❹ 烤盤刷上奶油，放上一層麵皮，鋪擺一層拌好的醬料。

❺ 重複疊放麵皮及醬料的步驟，直到材料用完，確定最上層為醬料，最後撒些佩可瑞若起司粉，並放幾塊奶油。

❻ 放進預熱至200℃（400℉）的烤箱中，烤25分鐘，直到表面金黃即可。

Preparation

❶ Roll out the dough into a thin sheet and cut into 8×12cm pieces for lasagne.

❷ Bring the salted water to the boil. Add the lasagna and cook for 4 minutes, stirring occasionally. Do not cook more than 7 pieces at a time. Drain with a sieve. Put in cold water and drain again. Arrange the lasagna one by one on trays, lined with dampened towel and covered with plastic wrap.

❸ Mix the pesto sauce with Bechamel and half the Pecorino cheese.

❹ Butter the baking dish. Drape the lasagna and spread over the sauce.

❺ Repeat the procedure until all the ingredients are used. Make sure the final layer is the sauce. Sprinkle with Pecorino cheese and some pieces of butter.

❻ Heat the oven to 200℃(400℉) and bake the lasagna for 25 minutes until the top is golden browned.

RAVIOLI DI SALMONE AL VERDE

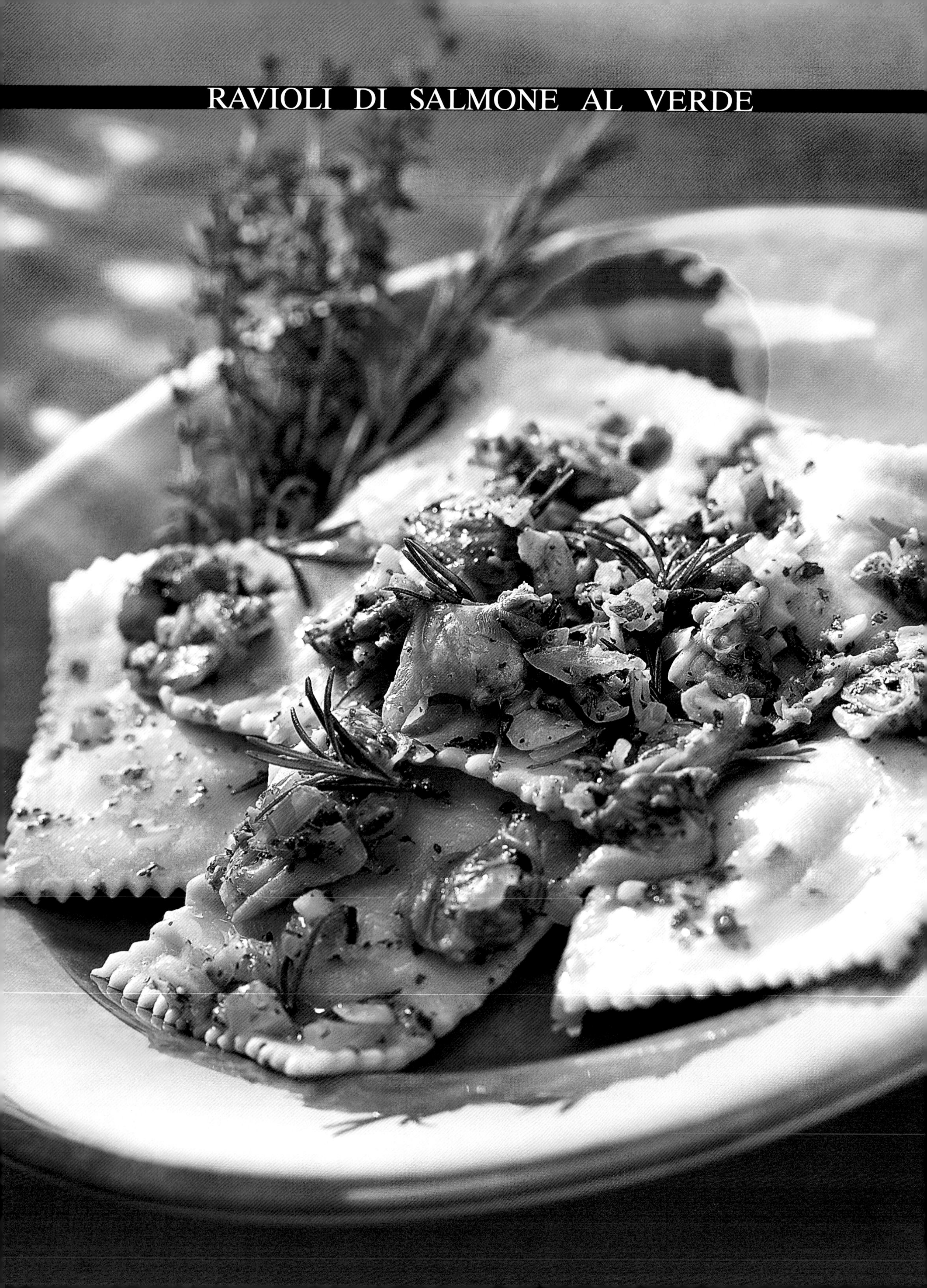

香料鮭魚餃

材料　　4～6人份

義大利麵團（見14頁）…	400公克
餡料部分：	
新鮮鮭魚	300公克
煙燻鮭魚	100公克
蛋	1個
鮮奶油	50公克
醬料部分：	
蛤蜊肉	300公克
洋蔥切碎	30公克
大蒜切碎	20公克
百里香切碎	10公克
鼠尾草切碎	10公克
迷迭香切碎	10公克
荷蘭芹切碎	10公克
特級純橄欖油	40公克
白酒	1/2杯
鹽、胡椒	適量

Ingredients Serves: 4-6

400 g pasta dough (see page 14)
For the stuffing:
300 g fresh salmon
100 g smoked salmon
1 egg
50 g whipping cream
For the sauce:
300 g shelled clams
30 g onions, chopped
20 g garlic, chopped
10 g thyme, chopped
10 g sage, chopped
10 g rosemary, chopped
10 g parsley, chopped
40 g extra-virgin olive oil
1/2 glass white wine
salt and freshly ground pepper

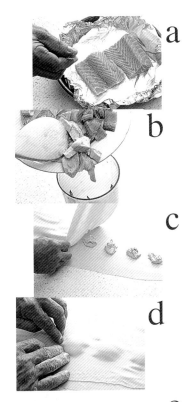

作法

1. 鮭魚放入烤箱，烤10分鐘，取出待冷卻。

2. 鮭魚肉、煙燻鮭魚、蛋、鮮奶油、鹽和胡椒放入調理機內，攪拌至濃稠狀。

3. 取一塊麵團擀成6至7公分寬的麵皮，在四周邊緣刷些水以便沾黏，以直線排列的方式，每隔3公分放1大匙（作法2）的餡料，再用另一張大小相同的麵皮覆蓋。

4. 壓出空氣，用滾輪刀將麵皮切成3公分見方的正方形，四周壓緊黏住。

5. 製作醬料：橄欖油入鍋加熱，放入洋蔥及大蒜，炒至洋蔥變軟。放入蛤蜊肉、所有香料、白酒、鹽和胡椒，煮到酒精完全揮發。

6. 另將魚餃放滾沸的鹽水中，煮6至8分鐘，煮時須稍加攪動，待浮上水面時，用漏杓撈起。

7. 將魚餃與醬料翻炒1分鐘，即可趁熱盛盤上桌。

Preparation

1. Bake the salmon in the oven for 10 minutes. Let cool.

2. Put the salmon, smoked salmon, egg, cream, salt and pepper in a food processor and mix until creamy.

3. Roll out a strip of dough into 6-7㎝ wide sheet. Brush the edges of each sheet with water for sticking. Place a tablespoon of fish mixture at 3㎝ intervals along the sheet. Cover with the second pasta sheet.

4. Press out the air pockets. With a pastry wheel, cut the pasta into 3×3㎝ squares. Press around each parcel to seal the edges.

5. For the sauce: Heat the olive oil in a frying pan. Add the onions and garlic. Sauté until the onions are softened. Add the clams, herbs, wine, salt and pepper. Stir until the wine has evaporated.

6. Meanwhile bring the salted water to the boil. Add the ravioli and cook for 6-8 minutes until al dente. Stir occasionally. Drain well with a sieve.

7. Toss the ravioli with the clams sauce and sauté for 1 minute. Serve immediately.

國家圖書館出版品預行編目資料

義大利麵家鄉風味／朱利安諾‧格薩里(Giuliano Gasali)
著；廖家威‧徐博宇攝.
--初版--臺北市：積木文化出版；城邦文化發行，民91
128面；21×28公分.（五味坊；26）（朱利安諾的廚房；6）
ISBN 978-957-469-933-1（平裝）
1.食譜-義大利　2.食譜-麵食
427.12　　　　　　　　　　　　　91000434

五　味　坊　**26**

朱利安諾的廚房　**06**

義大利麵家鄉風味（暢銷紀念版）

作　　　者／朱利安諾‧格薩里（Giuliano Gasali）
攝　　　影／徐博宇、廖家威
責 任 編 輯／陳嘉芬、何韋毅
外 文 審 校／吳凱琳

發 　行 　人／凃玉雲
總 　編 　輯／王秀婷
版　　　權／向艷宇
行 銷 業 務／黃明雪、陳志峰
出　　　版／積木文化
　　　　　　台北市104中山區民生東路二段141號5樓
　　　　　　電話：(02)25007696　　傳真：(02)25001953
　　　　　　官方部落格：http:// www.cubepress.com.tw
　　　　　　讀者服務信箱：service_cube@hmg.com.tw
發　　　行／英屬蓋曼群島商家庭傳媒股份有限公司城邦分公司
　　　　　　台北市民生東路二段141號2樓
　　　　　　讀者服務專線：(02)25007718-9　　24小時傳真專線：(02)25001990-1
　　　　　　服務時間：週一至週五上午09:30-12:00、下午13:30-17:00
　　　　　　郵撥：19863813　　戶名：書虫股份有限公司
　　　　　　網站：城邦讀書花園　網址：http://www.cite.com.tw
香港發行所／城邦（香港）出版集團有限公司
　　　　　　香港灣仔駱克道193號東超商業中心1樓
　　　　　　電話：852-25086231　　傳真：852-25789337
馬新發行所／城邦（馬、新）出版集團
　　　　　　Cite (M) Sdn. Bhd. (458372U)
　　　　　　11, Jalan 30D/146, Desa Tasik, Sungai Besi,
　　　　　　57000 Kuala Lumpur, Malaysia.
　　　　　　電話：603-90563833　傳真：603-90562833

封 面 設 計／呂宜靜
美 術 構 成／高櫻珊
製　　　版／上晴彩色印刷製版有限公司
印　　　刷／東海印刷事業股份有限公司

城邦讀書花園
www.cite.com.tw

2002年（民91）1月25日初版
2012年（民101）3月23日二版四刷　　　　　　Printed in Taiwan.
售價／350元

ISBN 978-957-469-933-1

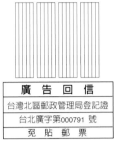

廣 告 回 信
台灣北區郵政管理局登記證
台北廣字第000791 號
免 貼 郵 票

積木文化

104 台北市民生東路二段141號2樓

英屬蓋曼群島商家庭傳媒股份有限公司城邦分公司　收

地址

姓名

請沿虛線摺下裝訂，謝謝！

積木文化

以有限資源・創無限可能

| 編號：VF0026X | 書名：朱利安諾的廚房 06—義大利麵家鄉風味（暢銷紀念版） |

積木文化　　讀者回函卡

積木以創建生活美學、為生活注入鮮活能量為主要出版精神。出版內容及形式著重文化和視覺交融的豐富性，出版品包括食譜、居家生活、飲食文化及家政類等，希望為讀者提供更精緻、寬廣的閱讀視野。
為了提升服務品質及更了解您的需要，請您詳細填寫本卡各欄寄回（免付郵資），我們將不定期寄上城邦集團最新的出版資訊。

1.　您從何處購買本書：＿＿＿＿＿縣市＿＿＿＿＿＿書店
　　□書展　□郵購　□網路書店　□其他＿＿＿＿＿＿＿＿＿＿＿＿＿
2.　您的性別：□男　□女　您的生日：＿＿＿＿年＿＿月＿＿日
　　您的電子信箱：＿＿＿＿＿＿＿＿＿＿＿＿＿＿＿＿＿＿＿＿＿
3.　您的教育程度：
　　1.□碩士及以上　2.□大專　3.□高中　4.□國中及以下
4.　您的職業：
　　1.□學生　2.□軍警/公教　3.□資訊業　4.□金融業　5.□大眾傳播　6.□服務業
　　7.□自由業　8.□銷售業　9.□製造業　10.□其他＿＿＿＿＿＿＿＿
5.　您習慣以何種方式購書：
　　1.□書店　2.□劃撥　3.□書展　4.□網路書店　5.□量販店　6.□其他＿＿＿＿＿＿
6.　您從何處得知本書出版：
　　1.□書店　2.□報紙/雜誌　3.□廣播　4.□電視　5.□其他＿＿＿＿＿＿＿
7.　您對本書的評價（請填代號 1非常滿意 2滿意 3尚可 4再改進）
　　書名＿＿＿　內容＿＿＿　封面設計＿＿＿　版面編排＿＿＿　實用性＿＿＿
8.　您選購食譜書最主要考慮的因素：(請依序填寫)
　　1.□作者　2.□主題　3.□攝影　4.□出版社　5.□價格　6.□實用　7.□其他＿＿＿＿＿＿
9.　您購買雜誌的主要考量因素：(請依序填寫)
　　1.□封面　2.□主題　3.□習慣閱讀　4.□優惠價格　5.□贈品　6.□頁數　7.□其他＿＿＿＿＿
10. 您學習烹調的優先順序為：(請依序填寫)
　　1.□中餐　2.□西餐　3.□烘焙　4.□日式料理　5.□其他＿＿＿＿＿＿
11. 您閱讀美食雜誌除食譜外最希望加入哪些單元？(請依序填寫)
　　1.□健康　2.□減肥　3.□美容　4.□飲食文化　5.□DIY餐具或廚房用品　6.□其他＿＿＿＿＿
12. 您最喜歡本書中的哪一道料理：
　　＿＿＿＿＿＿＿＿＿＿＿＿＿＿＿＿＿＿＿＿＿＿＿＿＿＿＿＿＿＿＿

　　原因　＿＿＿＿＿＿＿＿＿＿＿＿＿＿＿＿＿＿＿＿＿＿＿＿＿＿＿
13. 您最喜歡的食譜作者及出版社：

　　＿＿＿＿＿＿＿＿＿＿＿＿＿＿＿＿＿＿＿＿＿＿＿＿＿＿＿＿＿＿＿
14. 您希望我們未來出版何種主題的食譜書：

　　＿＿＿＿＿＿＿＿＿＿＿＿＿＿＿＿＿＿＿＿＿＿＿＿＿＿＿＿＿＿＿
15. 您對我們的建議：

　　＿＿＿＿＿＿＿＿＿＿＿＿＿＿＿＿＿＿＿＿＿＿＿＿＿＿＿＿＿＿＿